AQUACULTURE

AQUACULTURE

B. N. Pandey
Sadhana D. Pande
P.N. Pandey

A P H PUBLISHING CORPORATION
4435-36/7, ANSARI ROAD, DARYA GANJ
NEW DELHI-110 002

Published by
S.B. Nangia
A P H Publishing Corporation
7, Ansari Road, Daryaganj
New Delhi 110002
Ph.: 23274050
E-mail : aphbooks@gmail.com

2025

Printed at
Balaji Offset
Navin Shahdara, Delhi 110032

Dedicated to
Sri Sushil Kumar Modi
Hon'ble Deputy Chief Minister of Bihar

PREFACE

The present book entitled *"Aquaculture"* comprises 21 chapters by well known experts and research workers in their respective fields. We are thankful to all the contributors who responded promptly by making available their articles and it is the overwhelming response which has catalytic impact upon us to complete the task of publishing this book. Because of the overwhelming response from the contributors the proceedings of "17th All India Congress of Zoology and National Symposium on Co-existence with friendly fauna in India" held at Vidya Pratishthan's Arts, Science and Commerce College, Baramati, Dist. Pune (M.S.) held from 15th-17th October, 2006 was published in time.

We express our thanks to the Governing Council members specially Sri Sharadchandraji Pawar - President and Principal of the College Dr. Arun Adsool for their kind cooperation and blessings in making the Congress and Symposium a grand success and publishing the book well in time.

This book will be very useful for the students, teachers and researchers.

B.N. Pandey
Sadhana D. Pande
P.N. Pandey

CONTENTS

LIST OF CONTRIBUTORS

G.R. Bhraswadkar
Dept. of Zoology
Dr. Babasahed Ambedkar Marathwada University
Aurangabad (M.S.)

G.K. Kulkarni
Dept. of Zoology
Dr. Babasahed Ambedkar Marathwada University
Aurangabad (M.S.)

B.V. Yadhav
Dept. of Zoology
Dr. Babasahed Ambedkar Marathwada University
Aurangabad (M.S.)

D.B. Bhure
Dept. of Zoology
Dr. Babasahed Ambedkar Marathwada University
Aurangabad (M.S.)

Nitin Padwal
Dept. of Zoology
Dr. Babasahed Ambedkar Marathwada University
Aurangabad (M.S.)

A.A. Shaikh
Dept. of Zoology
A.C.S. College, Narayangaon (M.S.)

J.M. Jamadar
Dept. of Zoology
A.C.S. College, Narayangaon (M.S.)

A.R. Gulave
Dept. of Zoology
A.C.S. College,
Narayangaon (M.S.)

D.S. Darekar
Dept. of Zoology
A.C.S. College, Narayangaon (M.S.)

Khyade Vithalrao, B.
Dept. of Zoology
Shardaba Pawar Mahila Mahavidyalaya,
Shardanagar Malegaon (BK) Baramati (M.S.)

Ghantalus Uma. S.
Dept. of Zoology
Shardaba Pawar Mahila Mahavidyalaya,
Shardanagar Malegaon (BK) Baramati (M.S.)

Sinde Vandana, D.
Dept. of Zoology
Shardaba Pawar Mahila Mahavidyalaya,
Shardanagar Malegaon (BK) Baramati (M.S.)

Sonali S. Machale
Dept. of Zoology
Shardaba Pawar Mahila Mahavidyalaya,
Shardanagar Malegaon (BK)
Baramati (M.S.)

Uma, R. Ghantalu
Dept. of Zoology
Shardaba Pawar Mahila Mahavidyalaya,
Shardanagar Malegaon (BK) Baramati (M.S.)

R.G. Gaikwad
Dept. of Zoology
Shardaba Pawar Mahila Mahavidyalaya,
Shardanagar Malegaon (BK) Baramati (M.S.)

Kavita H. Nimbalkar
Dept. of Zoology
Shardaba Pawar Mahila Mahavidyalaya,
Shardanagar Malegaon (BK) Baramati (M.S.)

Ganga V. Mhamane
New English School,
Dhakale Tal - Baramati (Pune)

Vivekanand V. Khyade
Vidya Pratisthan College of Arts, Commerce and Science
Baramati, Dist. Pune

J.A. Kulkarni
Vidya Pratisthan College of Arts, Commerce and Science
Baramati, Dist. Pune

S.D. Deshpande
Vidya Pratisthan College of Arts, Commerce and Science
Baramati, Dist. Pune

R.B. Gaikwad
Vidya Pratisthan College of Arts, Commerce and Science
Baramati, Dist. Pune

V.V. Vaidya
Vidya Pratisthan College of Arts, Commerce and Science
Baramati, Dist. Pune

H.V. Mangi
Vidya Pratisthan College of Arts, Commerce and Science
Baramati, Dist. Pune

S.B. Hole
Vidya Pratisthan College of Arts, Commerce and Science
Baramati, Dist. Pune

K.J. Dhyansagar
Vidya Pratisthan College of Arts, Commerce and Science
Baramati, Dist. Pune

A.J. Yadhav
Vidya Pratisthan College of Arts, Commerce and Science
Baramati, Dist. Pune

Sneha G. Jagtap
Dept. of Zoology
Modern College, Pune

D. Dadel
Dept. of Zoology
Ranchi Women's College, Ranchi

L. Banerjee
Dept. of Zoology
Ranchi Women's College, Ranchi

P.N. Pandey
Dept. of Zoology
Ranchi University Ranchi

Seema Keshari
Dept. of Zoology
Ranchi University Ranchi

Purushottam Prasad
Dept. of Zoology
A.M.S. College, Barh, Patna

Smita Anand
Dept. of Zoology
A.M.S. College, Barh, Patna

P. Dubey
Dept. of Zoology
T.N.B. College, Bhagalpur

Mridula Banerjee
Dept. of Zoology
T.N.B. College, Bhagalpur

A. K. Poddar
Dept. of Zoology
T.N.B. College, Bhagalpur

N.B. Perveen
Dept. of Zoology,
Dhanajinana Mahavidyalaya,
Faizpur-425503 Jalgaon (M.S.)

Zeba Perveen
Dept. of Zoology,
Bibi Raza Women's College
Gulberga-585102 (K.S.)

S.G. Kulkarni
Dept. of Zoology,
Adarsh College, Omerga,
Dist. Osmanabad

S.S. Nanaware
Dept. of Zoology,
Yeshwant College,
Nanded (M.S.)

Arti Saxena
Dept. of Zoology,
Science College, Rewa (M.P.)

Usha Awasthi
Principal, Govt. Science College, Rewa (M.P.)

A.K. Tripathy
Dept. of Zoology,
Magadh University, Bodh Gaya

Shivani
Dept. of Zoology,
Magadh University, Bodh Gaya

1

Emerging R&D Issues in Aquaculture in Developing Countries in a Globalizing Economy

*VRP Sinha**

Fish captured from open water of oceans, seas, estuaries and rivers commonly come under the purview of fisheries and those farmed in controlled water, under aquaculture. Worldwide, aquaculture sector has grown at an average rate of 8.9 percent per year since 1 970, compared with only 1.2 percent for capture fisheries and 2.8 percent for terrestrial farmed meat-production systems over the same period. Aquaculture has increased almost 2.5 times during the last decade, reaching 31.2 million tonnes mostly coming from China, India, Indonesia and other developing countries, and it is expected that the global production could reach as much as 74 million tons by 2030.

Aquaculture sector is apparently very small compared to other food production sectors in volume but it significantly contributes to national income, nutritional security, export earning and in fulfilling social objectives, with a great promise to contribute further. Over a billion people meet their major source of animal protein from fish. However, there are emerging constraints in development of aquaculture, particularly because of the shortage of quality water, fish biodiversity, certain socio-economic issues and implication on fish habit and habitat of unexpected weather aberration because of global warming.

* 105 14 Forest Hill Drive, Wexford, PA 15090, USA

Fishery exports are essential to the economy of many developing countries. Presently, they accounted for about more than two-thirds of the total value of traded commodities in certain countries. Products derived from aquaculture production contribute an increasing share of total international trade in fishery commodities. Developed countries accounted for more than 80 percent of the value of total fishery product imports. Thus, because of globalizing economy many developing countries are gaining international markets and giving utmost priority to export of fish and fishery products.

Pragmatic research and development is needed for circumventing emerging constraints along with the matching government and NGOs support for infrastructures, credit and development of resources presently unsuitable for agriculture along with equitable distribution of quality water and water right for aquaculture. The paper highlights the areas of R&D urgently needed to meet the national requirements of fish and to gain and retain the international market by the developing countries in a globalizing economy.

INTRODUCTION

The global Green Revolution helped many developing countries to achieve self-sufficiency in food. But the future growth of agriculture does not seem to be that promising unless technological advances are made with spectacular results. Yields of major food crops have either reached their ceilings or have started to decline and FAO reported the expected shortfall of food crops in many countries. Dr. Diouf, the Director General. FAO noted recently that it is estimated that globally about 17 percent i.e. 815 million people, still suffer from hunger. However, the Green Revolution's original goal of 'productivity growth' now includes 'stability' and 'diversity' in food production. Fisheries sector in general and aquaculture in particular, may be a part of the solution to the increasing need for food diversity and arresting the decline in per caput nutrition.

On a global scale, cultured fish production has increased almost 2.5 times during last decade, reaching 31 .2 million tons, and output could reach 74 million tons by 2030. Fisheries sector besides being a source of cheap and nutritious food and of foreign exchange, is a powerful income and employment generator as it stimulates growth of a number of subsidiary industries. It is the source of livelihood for a large section of economically backward population of the developing countries.

Because of globalizing economy many developing countries have intensified export of fish and fishery products to earn foreign exchange either to pay off their debt of the international lending agencies or to meet the other essential imports.

The intensification of aquaculture in many developing countries has thus two major objectives, one to produce cheap protein food desperately needed for their population and the other to increase their share of international export of fish in a globalizing economy to earn foreign exchange. It is therefore highly important that the basic resources of aquaculture in terms of quality water, judicious utilization of primary productivity, protection of biodiversity, abatement of pollution of water, development of technology to utilize problematic soils through aquaculture and integration of aquaculture with livestock fanning and crop production, and sea farming are given highest priority. Matching policy research is required to provide support services to aquaculture development in terms of extension of technology, credit, marketing and processing infrastructural support at the national levels and safe guard the interests of aquaculturists at the international level through proper policy intervention and institutions.

AQUACULTURE BASIC RESOURCES AT STAKE

Shortage of Quality Water

Falling free from the sky for all, water has alluded everyone of its being plenty and an antidote of all the toxicity and also solvent of all the sin! But it needs to be realized by all that it is highly precious and has a tolerance limit of misuse and abuse. Since the inception of agriculture the great rivers — the Nile, the Tigris and the Euphrates, the great rivers of the Indian sub-continent and of China became the cradle of agriculture and in order to keep pace with the adequate food supply to growing human population, crop production and irrigation have been given the highest priority. Further, to control flood and availability of water in drought for irrigation, and for power generation, the water resources are modified so much so that it is said that ecologically, rivers are under siege. They are being drained, diverted, polluted, and blocked at a rate that has degraded freshwater ecosystem worldwide with more than half of the world's rivers stopped up by at least one large dam (Over 15 meters high) with more than 40,000 large dams bisecting waterways around the world and about 500,000 kilometers of rivers have been dredged and channelized for shipping (Runyan 2001). This development had benefited mostly the human population but unfortunately affected adversely fish and their habitat. Restriction on free flow of water current affects the oxygen content, temperature, sediment and organic material transport and other water qualities, adding problems of resident fish in adjusting themselves in the new environment. Many get perished in the process and some that are unable to adjust, eventually die off. Fisheries get affected and the fishermen particularly fishing downstream suffer most. The worst suffers are those who depended on the flood - plain fisheries.

The abstraction of water from rivers during summer months for crop cultivation puts serious limitation on fish biological activities because at that time many fish physiological requirements need more oxygen, and food and thus greater amount of water. Abstraction of water, in fact, is threatening the very survival of the ecosystem itself. The Aral Sea, a freshwater ecosystem, has become half in the last forty years with alarming adverse effect on its fisheries to the advantage of cotton cultivation in Uzbekistan. In China heavy water demands for agriculture has resulted in the dry lake beds on the Gianghan Plains where 1000 lakes existed in 1950 but only 300 remained after three decades. Similarly, Lake Chad in Africa is in serious ecological trouble. It has shrunk to one tenth of its former size. In 1960 it was 25,000 square kilometers, now it is down to 2,000 square kilometers. The lake fisheries have collapsed. (Hinrichsen 2003).

Unfortunately, it is not that the crop production and industrial production are taking major share of water but livestock farming has also emerged as one of the highest consumers of water. For example, one kg of grain-fed beef needs at least 15 cubic meter of water, and one kg of lamb from a sheep fed on grass needs 10 cubic meter compared to 0.4-3 cubic meters of water for a kg of cereals (Kirby, 2004).

To conclude, globally, water consumption has escalated to unprecedented limits and many parts of the world are showing signs of water stress with water table falling, lakes shrinking and wetland disappearing. Agricultural, industrial and urban demands have been on rise since 1950 putting pressure on ecological limits (Postel 1991). In. Asia, irrigation accounts for more than 90% of water withdrawals amounting about 1,210 billion cubic meter per year. Next is industrial consumption. However, water used by different sectors of economy, city dwellers, farmers and aquaculturists, returns back polluted in one form or other. The more the sector spends the more it pollutes the water. Thus, agriculture is not the major consumer of water but the major source of pollution as well because of heavy use of agro-chemicals.

Deterioration of Water Quality

Pollution of water is a serious problem. Increased population and production of livestock also pose serious environmental problems, since the fecal matter is a major source of nitrogen that is being added to the natural cycle. The present global population of animals is around 45 billion, which is expected to grow to 100 billion in another 20 years. In the USA about 1.6 billion tons of excrement is added to the nature from half of it feedlots. Raw pollution due to dairy production in France is equivalent to that of more than 250 million people (Kandel 2003). Nirenberg (2001) estimated that presently 2.5 billion pigs and cattle void more than 80 million tons of waste nitrogen annually compared to 30 million tons by the total human population.

Waste needs to be viewed as an important resource and a technology to use it hygienically for healthy fish production is needed. There is a growing awareness all over the world on the need to recycle sewage waste into protein wealth. Sewage water waste provides scope for large-scale culture of protein rich algae at the primary level and fish at the secondary treated level, which improves water quality, besides producing fish for food or fishmeal for animal feed. However, strict monitoring and surveillance is required to see that it should not become public heath *hazard.* Crop-livestock-fish culture technology needs up gradation to avoid any possible public heath problem.

Water related diseases, gastrointestinal infection, diarrhea are on the rise and it is estimated that some 600 million people in the world suffer from such infection, most of them in the developing countries resulting in death of about 20 million every year (Kandel 2003). Similarly, fish disease epidemics are on increase and will continue to be a threat to aquaculture development. The last two decades have seen the devastation caused by the Epizootic Ulcerative Disease Syndrome (EUS) in carp, catfish and many other fishes in wild and under culture conditions in different south Asian and Australian regions. Similarly, white spot disease has created havoc in coastal aquaculture particularly in shrimp farming all over Asia. Asian Development Bank and Network of Aquaculture Centers in Asia-Pacific in 1998 prepared a valuable report on Aquaculture sustainability and environment. It indicated that the disease problems are closely related to wafer quality and other environmental problems and showed a total loss of $104.6 million/yr. in carp culture in the region which is about half that in shrimp. Globally, the loss is estimated to be over $3 billion because of disease in shrimp aquaculture. Thorough knowledge of identification of pathogen, their mode of infestation, role of environment and of holistic pathology is essential to evolve methods of control of diseases. Health of fish and shrimps becomes a serious issue not only for production point of view but it has serious significance for public health as well as for export.

Almost every government has made laws regulating pollution but their implementation and public participation are weak and need effective steps for government intervention. However, it is highly important that Governments should ensure supply of proper quality water to aquaculture farms particularly in coastal area since estuarine water by and large is polluted with heavy metals and other pollutants mostly from industries. At the same time aquaculturists also need to discharge effluent water from farms after proper treatment.

Water management, whether in agriculture or in industry or in intensive aquaculture needs to be progressively based on recycling. The demand of water is staggering in intensive shrimp farming, for example, to maintain an ideal condition for shrimp growth, water exchange in the pond is essential which starts with 5% per day in the first month and thereafter of about 10% and about as much as 30% before harvesting the crop. In Taiwan to produce 4-11 tons of shrimps, the water requirement is about 11,000 m^3/t and for a production of 12-27t of shrimp about 29000-43000 m^3/t. Such heavy (consumption of water affects the ground water, and causes water table to fall and land subsidence becomes a common phenomenon. Water conservation and management are of vital importance for making shrimp fanning environmentally compatible (Sinha, 2002).

Loss of Fish Biodiversity

Construction of dams has seriously harmed the fish biodiversity. Besides, affecting the running water productivity and fish production, the decrease in freshwater flow increases salinity and favors marine predatory fish to invade the nursery areas in coastal water. The effect of Aswan High Dam on the coastal waters of Mediterranean is well documented (Drinkwater and Frank 1994). Further, reduction in nutrients transport to the sea reduced catches of Sardinella (Shuman 1995). It is reported that down stream of Aswan High Dam in Lake Nasser only now 25 species offish are found as compared to 75 fish species in 1940. Similarly, several fish species have been affected in the upstream of Lake Kariba in Zambia and Zimbabwe. While, *Clarius garepinus, Labeo* spp. and *Barbus* species declined, *Protopterus annectens* has become extinct. In Brazil in Parana River *Piaractus mesopotamicus* and *Brycon orbignymus* were eliminated.

Chinese Three George Dam recently completed is the largest hydroelectric dam in the world and it is expected that it will threaten existence of numerous fish species (Mc Cully 2001). Dams on East river a tributary of the Pearl river have seriously affected Chinese shad and *Cirrihinus molitorella*. Chenderoh Dam on the Perak River in Malaysia showed severe effects on Cyprinidae species.

Danube and Volga in Europe and Lower River Murray in Australia are showing decrease in fish biodiversity. In fact, 24 native fish species were lost in Aral Sea, the fourth largest freshwater lake in central Asia. Same is true in Pakistan in Indus and Jhelum rivers. In India 375 fish species in the river Ganges 46 are endangered, and of 127 fishes in the river Brahmaputra 11 are threatened as reported in 1984. It is a pity that the diversity of living organisms amounting to 5 to 10 million species of plants and animals as a result of 3 billion years of evolution involving mutation, recombination and natural selection is meeting such a fate (Khoshoo 1986).

Fishermen while collecting the wild seed of shrimp destroy millions of other fish seed in many parts of the world. Thus, indiscriminate catches of young ones of fish and shrimp result in iulable loss of biodiversity. Hilsa juvenile commonly known as Jatka is heavily fished in Bangladesh and India. About 3000 - 4000 tons of Jatka constituting 400-500 million are caught annually in Sandspur area of Meghna river in Bangladesh. Over 75 million aquatic organisms mostly fish and shrimp larvae are lost annually during wild collection of 1.5 million post larvae of *Panaeus monodon* for shrimp farming in Bangladesh. Similar situation prevails in neighboring West Bengal state of India. India loses 20 billion other fish and shrimp larvae for collecting 400 million wild shrimp annually for fanning. In Indonesia young ones hurl explosives and homemade bombs, simply made with a beer or water bottle filled with certain mixture of fertilizers and kerosene and a cheap underwater fuse, on school of fish resulting in massive damage of coral reef and mass killing of fish (Ryan 2001). Damage because of the bombs, sedimentation, cyanide fishing and coral mining have contributed to the degradation of 70% of the Indonesian reefs and have left only 6% in excellent condition. All these are going to adversely affect the biodiversity of fish, the very basis of future of fisheries and aquaculture development. It is highly essential for them to understand the implications of such acts on the very basis of their livelihood. Conservation of fish biodiversity, environmental sustainability, and effect of weather aberration because of global warming on fish habit and habitat are urgent aspects to be addressed by aquaculture scientists, planners and policy makers.

Underutilization of Primary Productivity and Eutrophication and loss of Productivity

Most of the developing countries have an adequate influx of sunlight throughout the year, which has the potentiality of a sustained rate of carbon assimilation as high as @about 4-8 g of biomass per meter square per day or about 30 tons of dry weight/ha/yr. Technologically, it is highly feasible to produce 1t/ha/yr of fish particularly herbivorous fish without feed or fertilizers but stocking the pond with proper size fingerlings in adequate quantity i.e., 5000/ha of 25 g in case of major Asiatic carp (they are expected to become 250 g in 8-10 months and with a maximum mortality of 20%). It does not need-highly specialized skill to undertake aquaculture for such a low rate of production except preparation of the pond and stocking of the fingerlings. But if the aquatic productivity is not optimally utilized through fish culture, it gradually disrupts the aquatic environment. With the increased nutrients run off to the aquatic ecosystem eutrophication sets in, which results in serious infestation of water hyacinth and other obnoxious weeds. The gradual appearance of the aquatic weeds from submersed variety to emergent variety shows highly distressed condition of the water resources. With the advent of the emergent weeds more and more silt is trapped and the quantity, quality and productivity of water are adversely affected. In India over 60-70% of ponds and oxbow lakes are infested with water hyacinth. Such situation exists in many developing countries and water hyacinth has become one of the dominant factors for stressing and disrupting the aquatic ecosystem. Two parents plants of water hyacinth produce about 30 offspring in 23 days resulting in 12000 plants, with a total wet weight of 470 t/ha/ 4 months, while registering 5% gain in weight every day (Barrett 1989, quoted in Sinha 1999). In addition to out-growing other aquatic plants and animals, its growth results in trapping silt in large quantities and the loss of water through transpiration and plant growth, which adversely affects the carrying capacity of the ecosystem. Since water hyacinth contains 95-97% of water, it amounts to a potential loss of 446 tons of water/ha in 4 months, which is certainly most undesirable. Fish is a major ecological player and aquaculture needs to be viewed as a tool to utilize primary productivity to help maintain water and soil quality in the pond.

Loss of Coastal Aquaculture Resources

During the last six decades, large tracts of mangrove forests were cleared for fuel wood, paddy cultivation, saltpans and human settlement and for shrimp farming in many Southeast Asian counties and thus they lost considerable area of their mangrove. Large-scale destruction of mangrove has a number of ecological implications, since it provides natural breeding habitats and acts as nurseries for a number of marine and freshwater fish and prawn. Apart from soil erosion and subsequent increase in sediment load, the loss of forest cover upset the natural balance of mangrove ecosystem and adds to the fury of cyclones in the area.

Coastal eco-system is further stressed because of growing use of pesticides, insecticides and herbicide in agriculture, and because of increasing discharge of industrial effluent and human and animal waste. Similarly, industrial effluents also go to a larger extent to the various bays, creeks, and localized open shore regions of many countries which have direct adverse effect on feeding, breeding and nursery grounds of many fin and shellfish, ultimately affecting their marine fisheries. It is important to assess the damage being done to the coastal ecosystem and to evolve scientific guidelines for its rehabilitation.

To conclude, the situation will be much more serious in future in relation to all basic resources. According to Swaminathan (1996) the availability of renewable resources such as fish catch, irrigated land, cropland, water and forest compared to population size will show negative per capita availability by 2010. This will result in severe socio-economic problems and will also threaten environmental sustainability.

Urgent need of Pond Improvement and Farm Design

Ponds in many countries of Asia. Africa and Latin America are in defunct state. In many countries there are many reasons of apathy towards pond improvement. Seldom the pond belongs to an individual, most of the times it has multiple ownership. Individual ownership of the parent over the time got multiplied among the children. A joint decision for investment on pond improvement becomes difficult because of varying interest of the partners. Certain countries adopted policies for state control on such properties, which have not been very successful because of many reasons. Even after hundreds of years of such traditional aquaculture practices, which improved through many scientific innovations over the years, construction of ponds and farms remain still primitive and lack proper engineering design. The situation is very serious in brackish water aquaculture where water, silt, and effluent management becomes very difficult and directly affects the productivity. The situation is aggravated more because of frequent cyclones and mangrove deforestation. The scientific design should provide mangrove belt along the embankment.

Aquaculture engineering ingenuity is required to design efficient pond and farm keeping in mind the minimal loss of water through seepage and evaporation, water structure to control flooding, recirculation of water with provision of biological filter facilities for most efficient water management silt management devices and for treatment for effluents.

Management and recycling of the nutrients from pond humus is very important, since it is a potential manure with long lasting fertility effect. One ton of pond humus has fertilizing effect equivalent of 6 kg of ammonium sulfate. It is also estimated that a properly managed pond produces over 50 tons of humus and silt/ha/yr. Under utilization of humus results in deterioration of condition of the pond soil and consequently anaerobic state sets in the pond bottom affecting adversely the productivity.

SUGGESTIONS TO CREATE ADDITIONAL RESOURCE BASE FOR AQUACULTURE

Emphasis on utilization of agriculturally unsuitable areas for aquaculture

Utilization of problematic soils such as water logged, saline, alkaline and acidic normally unsuitable for agriculture can be profitably utilized by subsistence level aquaculture and that will add to the basic resources of aquaculture. For example, India has about 8.5 million ha of water logged and saline areas. In addition to these 13.6 million hectares of wasteland is available. Aquaculture technologies to use these soils already exist. Successful experiments under All India Coordinated Project on Composite Fish Culture repeatedly have shown Asiatic carp production@ 1.5 t/h/yr in Sunkesula Fish Farm, Karnool in A.P. in highly alkaline soil. Similarly, fish production under the project showed about 3-4 t/ha/yr of carp production in highly acidic soil in Gwahati, Assam (Sinha and Ramchandran 1985). Upland saline lakes and saline soils can be profitably used through culture of brackish-water fish and shrimps as has been shown by the Central Institute of Fisheries Education in Haryana in culturing shrimp and milk fish. The developing countries should view these resources as an additional aquaculture resources and utilize them for aquaculture. Similarly, paddy-cum-fish culture for concurrent and alternate cultivation of fish and rice and raising of fish seed in paddy field should be intensified wherever feasible.

Water right to aquaculturists and right to water

Within the framework of integrated resource management, requirement of water for aquaculture needs to be considered in land use planning, coastal zone planning and water use allotment particularly in water conservancy programs. Formulation of legal framework for water rights along with rights to water based resources is necessary to allow access and user rights to aquaculturists to farm fish in water bodies which are by and large under public domain in many countries. In this regards it is important to note that there exists possibilities of large-scale utilization of canal water for environmentally compatible cage and pen culture of fish.

Water rights in the Western world have been invariably centered on navigation. Ward (2002) noted that it was not until 1973 that Colorado in the USA put in place a law that gives guarantee of water for fish and wildlife. Such law needs to in place in developing countries where water resources are fastly modified ignoring the fish habit and habitat.

EMERGING SOCIO-ECONOMIC R&D ISSUES.

Generally it is seen that the price spread from farmers to consumers is at least about three times or even more in many countries. This is mainly because of the middlemen and marketing cost. The marketing system is more urban oriented than rural oriented. The wholesale channel members are often urban based and make more profits than fish farmers. Major share of marketing profits goes to urban sector while rural sector mostly receives its labor charges. Attempts to improve upon this through co-operatives met with varying degree of success and sometime failed due to lack of expertise, experience and resources with the cooperatives. It is important to undertake study to reduce the marketing cost so that producers get more profit and consumers pay less.

Presently, most of the benefits go to the large fish farmers or entrepreneurs who are generally city dwellers, often alien to the area. They get the leased land of small farmers to operate aquaculture enterprises. This leads to coastal tension and deprives small farmers of the benefit of coastal aquaculture. Proper studies need to be undertaken for policy formulation for government intervention.

Conflict between farmers and aquaculturists

Many countries have been experiencing serious conflicts between paddy grower and aquaculturists because of the conversion of paddy land to aquaculture, particularly in view of prevalent land tenure system and user rights in certain areas.

This needs in-depth study for policy formulation for government intervention to avoid sectoral conflicts.

Social issues in women participation in aquaculture

Women in Asian countries are involved in aquaculture production - whether secondary to their men as in most of

Asia, or as their equals as in China after 1949. Women also take part in inland fishing, in open lakes, rivers and streams. In Laos, they fish in canals. In the Philippines, they fish from canoes in coastal lagoons. In many African countries, they fish the rivers and ponds. Yet, their contribution never got the appreciation and support as required in improving their social status and acquiring empowerment. More women than men remained still malnourished and in many developing countries pregnancy and childbirth still remained the largest single cause of death in fishing communities. Wages of women are 30-40% less than men for comparable work.

Several international aid agencies recognized that the failure of many of their development projects was due to the exclusion of women in the design and implementation of the projects. Following the Proclamation of the UN Decade of Women in 1975, a number of major initiatives were launched by the FAO and other international organizations to draw attention to the contribution of women in rural development and agriculture, including fisheries. Many international and regional agencies with the national governments and the NGOs started initiatives on improving the social status of women, particularly those in small-scale fishing communities, by enabling them to participate in development and to engage in economic activities on equal terms with men with varying degrees of success. They are also increasingly assisting in many ways for involvement of women in aquaculture activities. Many countries have initiated special financing; the most successful is of micro-credit for fisheries and aquaculture for women. However, there is a serious need of evaluation of impacts of such programmes on women nutritional status, empowerment and exploitation from their counterparts.

There is an emergence of a new class of House Husbands where it is seen that women increased participation in fisheries or aquaculture activities have led their male counterparts to become more lethargic and confined at home activities and resort to maximum liquor consumption. This results in further exploitation of women. This has been more prevalent where females were going actively for the collection of shrimp post larvae. One of the first actions needed to redress the situation is to increase awareness of gender issues and to dispel perceptions that women are weak and helpless.

In this regards, it is interesting to note Madeline Albright, a former U.S. Secretary of State, that the biggest challenge to the 21st Century will undoubtedly be the conferment of basic human rights to women and of all the forces that will shape the world, the movement to recognize and realize the rights of women will be the most powerful.

INTERNATIONAL TRADE OF AQUACULTURE PRODUCTS

Fishery exports are essential to the economy of many developing countries. Presently they accounted for about more than two-thirds of the total value of traded commodities in certain countries. Products derived from aquaculture production contribute an increasing share of total international trade in fishery imports. Japan continued to be the largest importer of fishery products accounting for some 23 percent of the global trade and its fishery imports accounted for 4 percent of its total merchandise trade. The EC is also heavily dependent on imports of fish and fishery products. The United States, the world's fourth largest exporting country, was the second largest importer (FAO 2000). The net receipts of foreign exchange by developing countries increased from US$ 3.7 billion in 1980 to US$ 22 billion in 2001 – a more than 2.5-fold increase in real term. This increase was greater than the net exports of other agricultural commodities such as rice, coffee, and tea in many developing countries. Thus, because of globalizing economy many developing countries are gaining international markets and giving utmost priority to export of fish and fishery products to earn foreign exchange.

Many governments try to exploit as much of the basic resources as it can for exports and thus attach greater importance to export even at the cost of depriving their national of cheap protein food to fight existing serious malnutrition. As is well known many countries have serious malnutrition problems, for example, majority of Bangladeshi children suffer with malnutrition and over 30,000 children suffer with blindness annually and infant mortality is of very high order (World Bank 1998). Similarly, 200,000 Peruvians under the age of five are suffering from malnutrition and almost half of the pregnant women are suffering from anemia. Such is the case in many Asian, African and Latin American countries. Many experts contend that the exports of seafood, adversely affected the fisher folk. Globalization and the pressure to increase exports have devastated the livelihoods of tens of thousands of them. Multinationals entering the country under liberalized trade rules have also evicted existing populations from their coastal habitat, and set up polluting industries that discharge harmful effluents and cause health problems to the inhabitants there.

The large seafood export processing industry has provided new opportunities for employment in certain areas. However, this has also affected the availability of low quality fish for poor communities. Improved transportation and infrastructure have reduced village-level processing activities in many countries. The introduction of net making machines has displaced large numbers of women from net making. One factory near Nagercoil in India displaced at least 1000 women in the Kanyakumari district in 1980.

Article VI of GATS 2000 (General Agreement on Trade and Services) has been criticized as violating fundamental social rights enshrined in the UN's Universal Declaration of Human Rights and its accompanying charters. It is difficult to comment on the export policy of the governments because it is made on countries national priorities, compulsions and necessities. But it should be admitted also that many developing countries are striving hard with the help of donors and NGOs in many ways to brine respite to poor fishermen and women with varying degree of success.

Fish market infrastructure facilities in many countries are far from satisfactory or yet to be properly evolved. There is a strong need to create market infrastructure along with supply centers, and better credit facilities to avoid middlemen. Marketing cost of aquaculture commodity is very high, which needs to be minimized, through proper investment in improved transportation infrastructure. Studies need to be undertaken to formulate the policy guidelines for government intervention.

There is an acute shortage of skilled fish farmers in many countries where there was no tradition of aquaculture. Study needs to be undertaken to assess the farmers' training needs, their present skill and aptitude to embark on their training in aquaculture activities.

RESEARCH INSTITUTIONAL WEAKNESS

Lack of cost-effective mechanism of reliable data collection and lack of firm socio-economic indicators to assess the progressive socio-economic upliftment of rural poor with adoption of any technology or government policy adoption and implementation are great impediments in assessing the impacts and also in convincing planners and policy makers in many developing countries about the quantified role of the sector to get the required political, financial and public required support.

At present Fisheries and Aquaculture institutions in developing countries are to a great extent involved in biological and ecological studies involving disciplines of Taxonomy, Ecology, Diseases, Genetics, Physiology, Reproduction, Endocrinology, Nutrition, Microbiology, Fish Behavior, Biochemistry etc.

Some institutions have expertise on Aquatic Ecology, Aquatic Chemistry, and Aquaculture Engineering etc. for ecological intervention. However, for socio-economic management, which needs expertise on Economics, Sociology, Food technology, Marketing and Management Science are almost non-existent. In the present context of globalization, they become an urgent necessity to study different socio-economic issues and formulate policy guideline for government timely interventions at the national and international levels.

GENERAL REMARKS

With the opening of the global markets, the developing countries have invested heavily in the infrastructures for fisheries and aquaculture to increase export of fish and fishery products and they have been successful in their efforts in varying degree. However, at times, ill planned development resulted in over-exploitation of resources, serious deterioration of water quality, fish diseases and loss. The developing and least developed countries are likely to suffer in future in export of fish and fishery products because of global market distortion caused by subsidies, tariffs and other technical barrier to trade in order to discourage farmers in developing countries from becoming competitive, obviously because about 80% (by value term) of import is done by developed nations. Use of antibiotics, over-exploitation and loss of biodiversity become a serious public concern in developed countries affecting export. A law on traceability of fish, requiring producers to indicate commercial and scientific name of the species, its origin, methods of production wild or cultured, will put an extra strain on the fishing communities which are by and large not that educated in developing countries. It is essential that the governments should urgently move to motivate the fishermen aquaculturists and processors to meet with these requirements in order to gain and retain their international markets.

REFERENCES

Drinkwater, K.F. and K.T. Frank (1994) Effects of river regulation and diversion in marine fish and invertebrate. Aquatic Conservation, Freshwater and Marine Ecosystem 4:135-151.

FAO. 2000. The Status of World Fisheries and Aquaculture 2000. Fisheries Department. FAO of the United Nations, Rome, Italy.

Hinrichsen, D. (2003). A Human Thrust. World Watch. Jan/Feb. 2003 p. 15.

Khoshoo. T.N. Role of Science and Technology in Environmental management. The shaping of Indian Science. Vol. III : 1982-2003. pp 1387-1566 University Press (India) Pannal Limited, 3-5-819 Hyderabad, Hyderabad-500 029.

Kendal R. (2003) *Water from Heaven,* Columbia University, Press, New York. p. 312.

Kirby, A. (2004), Hungry World 'Must Eat Less Meat'. BBC. News Online 15th August 2004. UK.

Mc Cully, P. (2001) Stream of Consciousness, *World Watch.* Jan.Feb. 2001; 36-39.

Mc Ardle, J and A.M. Bullock. (1987) Solar ultraviolet radiation as a causal factor of 'Summer Syndrome' in cage reared Atlantic salmon, *Salmo salar* L, A clinical and histopathological study. *Journal of Fish Disease,* 1987(10): 255-264.

Nirenberg. D. (2001) Toxic fertility. *World Watch,* Vol 14(2): 30-38, March April 2001.

Postal, S. (1991) *Oasis, World Watch Series,* Washington

Runyan, C. (2000). All the Wild Rivers, *World Watch.* Jan.Feb. 2001; 31-33.

Ryan J.C. (2001). Indonesian's Coral reef on the Line, *World Watch,* 14(3): 13-19. May-June 2001.

Shuman, J.R. (1995). Environmental considerations for assessing dam removal alternatives for river restoration. Regulated Rivers: Research and Management 11:249-261.

Sinha, V.R.P. (1994). Introductory remarks on sustainable development of marine fisheries resources in Bangladesh: *Proc. of the Seminar on Sustainable Development of Marine Fisheries Resources in Bangladesh,* (Ed. By M.A. Mazid, V.R.P. Sinha and Md. Kamal). BFRI and FAO.

Sinha, V.R.P. and V. Ramchandran (1985) *Freshwater Fish Culture ICAR,* New Delhi pp. 80.

Sinha, V.R.P. (2002). Aquaculture and Aquatic Environment Sustainability. *The Sixth Indian Fisheries Forum,* 17-20 Dec., 2002. pp. 25-32, CIFE, Mumbai, India.

Sinha, V.R.P. (2005) *Fisheries Research Planning and Management in Developing Countries,* NPH, Publication 199, pp. Delhi.

Swaminathan. M.S. (1996) Towards an Ever-green Revolution. Souvenir, 2nd International Crop Science Congress, 17-24 November 1996, New Delhi, India pp. 29-35.

Ward. D.A. (2002) *Water Wars - Drought, Flood. Folly, and the Politics of Thirst..* Riverhead Books, New York, 294 pp.

World Bank (1998) *Bangladesh 2020 - A Long-run Perspective Study,* The World Bank and Bangladesh Centre for Advanced Studies. The University Press, Dhaka pp. 123.

World. Watch (2002) Environmental Intelligence, World. Watch March/April 2002:8.

Evaluation of Forms of Diet on Growth, Survival and Feed Utilization of Deccan Mahseer, *Tor Khudree* (Sykes) Fry

PATHARE, G.S., INDULKAR. S.T. AND PATIL, S.D.

SUMMARY

Effect of three forms of diet such as ball, pellet and flake was observed on growth, survival and feed utilization of Mahseer *(Tor khudree)* fry. Diets were prepared using locally available low cost ingredients such as fish meat powder, shrimp meal, acetes meal, groundnut oil cake, rice bran and wheat flour in equal proportions and fed to the Mahseer fry (22.08 ± 1.4 mg in weight and 15.4 ±0.34 mm in length) for 30 days. The experiment was conducted in 40 L plastic tubs following standard statistical method. Based on the observations made, higher weight gain (108.14%), length gain (21.15%), specific growth rate (2.41), protein efficiency ratio (0.56), survival (89.29%) and better food conversion ratio (5.81) was observed in fry fed on flake feed. Flake feed gave 40.22 per cent increment in weight gain and 22.15 per cent more survival than the ball feed. However, there was no significant difference ($p > 0.05$) in growth and survival of Mahseer fry fed on flake and pellet forms.

INTRODUCTION

Mahseer, the largest of the freshwater scaly fish, are commercially important game and food fish of India. According to ichthyologist Mahseer come under threaded fishes due to

S.D. College of Fisheries (Dr. B.S. Konkan Agricultural University), Ratanagiri - 415 629 (MS), India indulkarst@rediffmail.com its decline catch. Several species of Mahseer inhabit rivers and streams at an altitude from few metres to 2000 metre above the mean sea level, tolerating water temp. in the range of 6 °C to 35 °C. Mahseer have been of considerable importance because of their large size and durability after capture. It is necessary, therefore, that this valuable resource is properly managed, conserved and propagated both for food and sport.

Mahseer, well known game fish has fortunately caught the imagination of scientist who fears its extinction. Dr. C. V Kulkarni Ex-Fisheries Director, Maharashtra and Advisor to Mahseer Project of Tata Electric Company's Fish Farm, Walwhan near Lonavala, Pune had worked on breeding, seed production, conservation and propagation of Mahseer and reported in his book "Conservation of the Mighty Mahseer of India" that, one species *Mussulah* Mahseer, is already suspected to have passed into oblivision. The National Commission on Agriculture (1976) in their report on fisheries stated, "there is general decline in Mahseer fishery due to indiscriminate fishing of brood and juvenile fish and the adverse effect of river valley projects. Therefore, they have suggested for extensive surveys and detailed ecological and biological investigations of this important group of fish. Conservation of Mahseer is now put forward by many environmentalists and angler's societies. Conservation practices include ranching, declaring sanctuaries, breeding and nursery rearing of spawn, fry and fingerlings for stocking in the reservoirs and ponds (Kulkarni and Ogale, 1995).

In hatchery practices, larval rearing is critical phase in which maximum growth and realisation are expected. Proper feed and good water quality management can achieve more survival during nursery rearing. In any larval rearing practices, feed plays a vital role to get healthy fry and fingerlings. Therefore, it is necessary to know the food and feeding pattern of any cultivable species. The trial results of fry to semi fingerlings rearing under cold climatic conditions of Lonavala for 13 days have indicated only 19 per cent survival (Shirgur and Indulkar, 1994). The present method of nursery rearing of Mahseer *(Tor khudree)* at Lonavala farm resulted in 30 to 40 per cent survival (Personal communication). This is the only centre in the state where breeding, nursery rearing and ranching of seed are being undertaken every year for conservation and propagation of this important game fish. At this centre, for nursery rearing of Mahseer fry, the combination of fish meal, shrimp meal, acetes meal, groundnut oil cake, rice bran and wheat flour in equal proportion in wet ball form is being used as feed. The use of feed in wet ball form has some drawbacks such as, poor survival of fry, wastage of food, maintenance of water quality etc. Thus, to propagate culture their by conservation of this important game and food fish, it is very essential to improve the nursery rearing practices. The reports available on development of particulate diet for maximum growth and survival at nursery rearing of *T. khudree* are limited. With this aim, present study was planned to see the effect of different forms of diet on growth and survival of Deccan Mahseer fry *(Tor khudree)* during indoor nursery rearing.

MATERIALS AND METHODS

In the present study diet was prepared in three forms, viz., ball, pellet and flake. Per cent composition of ingredients and proximate analysis of diets prepared for experiment are given in Table 2.1.

Table 2.1: Percent composition of ingredients and proximate composition of various forms of diet.

Particulars		*Test Diets*	
Ingredients (%)	T_0	T_1	T_2
Fish meat powder	16.66	16.66	16.66
Prawn head meal	16.66	16.66	16.66
Acetes meal	16.66	16.66	16.66
Groundnut oil cake	16.66	16.66	16.66
Rice bran	16.66	16.66	16.66
Wheat flour	16.66	16.66	16.66
Vitamin (Tablets 100 g^{-1} of feed)	1	1	1
Proximate Composition (per cent on dry weight basis)			
Moisture	15.4	6.0	5.8
Crude Protein	31.4	32.6	32.8
Crude Fat	3.4	8.6	9.2
Total Ash	16.8	13.8	14.2

1. *Ball (T_0) control diet* : The required quantity of grinded ingredients were mixed in mixer. Adequate quantity of water was added to make a dough. Then, dough was steamed in a cooker for 10-15 minutes. A vitamin tablet was added to cool dough and wet balls were prepared.
2. *Pellet (T_i):* Dough was prepared as above and steamed it in a cooker for 10 to 15 minutes. Vitamin tablet was added in cooled dough and thoroughly homogenized. The cooled dough is passed through extruded fixed with 1.0 mm diameter die and pellets were prepared. The prepared pellets were sun dried and stored.
3. *Flake (T_2):* The dried ingredients in powder form were passed through 0.5 mm mesh size and were weighed as per the required quantities and mixed thoroughly in mixer. Required quantity of water was added in mixture. Then the mixture was steamed till it form a slurry. Vitamin tablets were added in cooled slurry then this slurry was spread on a black polythene sheet with the help of smooth brush in a layer of one mm thickness and was kept for sun drying. The sun dried flakes were removed from the polythene sheets and kept in oven at 60 to 70 °C temperature for 30 minutes to remove excess moisture.

The experiment was conducted in 40 L capacity plastic tubs keeping 30 L water in each tub. Before starting the experiment, a group of 30 numbers of fry chosen randomly and there initial weight and length were recorded. The volume of water was maintained at 30 L in each tub during entire period of experiment. Fry were stocked at the rate of ten fry per tub in all the three experiments i.e. one fry per three liters of water. Three treatments were carried out during this experiment. For each treatment, seven replicates were carried out. Feeding was done at the rate of 20 per cent of collective initial body weight of fry. The calculated feed was divided into two parts and given during morning (08:00 hours) and evening (18:00 hours). The left over feed and excreta of fry were removed daily. Nearly 25 per cent water from each container was replaced after each siphoning. The experiment was conducted in the laboratory at room temperature for 30 days of rearing period.

During the experimental period routine water parameters, such as, water temperature, pH, alkalinity and dissolved oxygen were recorded by adopting standard methods (APHA, 1998).

At the end of 30 days of rearing period of the experiment, the fry were counted from each replicate and their individual length and weight were recorded. The average value of weight and length were taken from each replicate of a treatment for analysis of growth parameters. Standard methods were adopted to find out the growth parameters, such as body weight, body length, gain in body weight, gain in body length, per cent gain in body weight, per cent gain in body length, specific growth rate, protein efficiency ratio, food conversion ratio and survival (Chow, 1980).

The mean weight and length of the harvested fry from each replicate of treatment was calculated and served as one replicate. Standard statistical methods were adopted for testing significance. One-way analysis of variance (ANOVA) was applied to find out the difference among the mean value of growth and survival of fry. Significant difference was indicated at 0.05 level. The Newman Keuls' multiple comparison test was used to find out the difference among the treatments (Zar, 1974).

RESULT AND DISCUSSION

Observations on growth and survival of fry fed on different forms of diet for 30 days are given in Table 2.2.

The results indicated that percentage weight gain of Mahseer *(Tor khudree)* fry was 67.92, 99.40 and 108.14 for treatments T_0, T_1 and T_2 respectively. ANOVA showed significant difference ($P<0.05$) in per cent weight gain of fry fed on different forms of diet. Significantly minimum percentage weight gain was observed in T_0 (67.92%) whereas, T_2 fed fry showed higher percentage weight gain (108.14%). Newman Keuls multiple comparison test (SNK) showed that, there was no significant difference ($P>0.05$) in per cent weight gain of fry fed on T_1 and T_2, whereas, weight gain (%) of Mahseer fry fed on T_0 differed significantly ($P<0.05$) with treatments, T_1 and T_2.

Table 2.2: Growth and Survival of Fry *Tor khudree* Fed on Different Forms of Diet

Particulars	*Diets*		
	To	T_1	T_2
Initial average weight (mg)	22.08 ±1.40	22.08 ±1.40	22.08 ±1.40
Initial average length (mm)	15.40 ±0.34	15.40 ±0.34	15.40 ±0.34
Weight (mg) after 30 days			
R.	35.60	43.89	59.43
R_2	34.60	44.23	46.83
R_3	43.33	43.92	37.43
R4	38.60	44.29	37.5
R_5	38.83	43.83	46.00
R6	35.70	43.98	46.71
R_7	32.87	44.06	47.8
Mean	37.08 ±1.31	44.03 ± 0.07	45.96 ±2.81
Length (mm) after 30 days			
Ri	16.80	18.50	19.40
R_2	16.70	18.10	18.40
R_3	17.20	18.30	18.60
R4	16.60	18.80	18.10
R_5	17.40	18.40	18.20
Re	16.70	17.90	18.80
R_7	17.20	18.60	19.10
Mean	16.94±0.12	18.37±0.11	18.66±0.18
Gain in Weight (mg)	15.00±1.31	21.95±0.07	23.88±2.81
Gain in Length (mm)	01.54±0.12	02.97±0.11	03.26±0.18
Average percentage gain in Weight	67.92^a±5.95	99.40^b±0.30	108.14^b±12.70
Average percentage gain in Length	10.02^a±0.77	19.29^b±0.75	21.15^b±1.17
Specific growth rate (% day^1)	1.72^a± 0.12	2.30^b±0.005	2.41^b±0.20
Food conversion ratio	8.21^b±0.51	5.82^a±0.04	5.81^a±0.61
Protein efficiency ratio	0.40^a±0.03	0.53^b±0.004	0.56^b±0.06
Survival (%)	67.14^a±7.14	65.71^a±3.69	89.29^b±3.69

a, b : Value in a row sharing similar letter do not differ (P>0.05).

± : Standard Error of mean.

The percentage length gain of fry in treatments T_0, T_1 and t_2 was 10.02, 19.29 and 21.15 respectively. ANOVA showed significant difference (P<.05) in percentage length gain of fry fed on different forms of diet. The minimum and maximum percentage length gain was observed in fry fed on TO (10.02) and T_2 (21.15) respectively. SNK showed that T_1 and T_2 did not differ significantly (P>0.05) whereas, treatment TO differed significantly from treatment T_1 and T_2.

The specific growth rate (SGR) of fry fed on different forms of diets for 30 days was 1.72, 2.30 and 2.41 % day^{-1} for treatment T_0, T_1 and T_2 respectively. ANOVA of SGR indicated significant difference (P<0.05) among the diet forms. The maximum SGR was observed in fry fed on T_2 (2.41% day^{-1}) whereas, minimum SGR was observed in fry fed on TO (1.72% day^{-1}). By computing SNK, it was observed that there was no significant difference (P>0.05) in SGR of fry fed on TI and T_2 diet forms.

The food conversion ratio (FCR) of fry was 8.21, 5.82 and 5.81 for treatment to, T_1 and T_2 respectively. The minimum FCR was found in fry fed on T_2 (5.81) and maximum in T_0 (8.21) diet. ANOVA showed significant difference (PO.05) in FCR of fry fed on different forms of diet. SNK indicated that there was no significant difference (P>0.05) in FCR of fry among treatments T_0 and T_2.

The protein efficiency ratio (PER) was 0.40, 0.53 and 0.56 in treatment T_0, T_1 and T_2 respectively. The maximum PER in fry was obtained for the diet T_2 (0.56) and minimum in diet T_0 (0.40). ANOVA showed significant difference (P<0.05) in PER of fry fed on different forms of diet. However, by computing SNK it was observed that, there was no significant difference (P>0.05) in PER of fry fed on TI and Ta. The PER in treatment TO has significantly differed (P< 0.05) from treatments T_1 and T_2.

The survival (%) of fry after 30 days of rearing period was 67.14, 65.71 and 84.29 in treatment T_0, T_1 and T_2 respectively. Maximum survival (%) was found in T_2 (84.29) and minimum in T_1 (65.71). ANOVA showed significant difference (P>0.05) in survival of fry fed on different forms of diet. Survival of fry in treatment T_1 has significantly differed (P>0.05) from treatment T_0 and T_1

During the experimental period water parameters such as, temperature dissolved oxygen, and pH was recorded weekly. In treatment T_0 temperature varied in the range of 24 to 27 °C, pH varied in the range of 8.0 to 8.2, dissolved oxygen in the range 8.2 to 8.8 mg L^{-1} and total alkalinity varied in the range of 220 to 225 mg L^{-1}. In treatment T_1 temperature varied in the range of 24 to 27 °C, pH varied in the range of 8.2 to 8.4, dissolved oxygen in the range 8.3 to 8.8 mg L^{-1} and total alkalinity varied in the range of 221 to 228 mg L^{-1}. In treatment T_2 temperature varied in the range of 24 to 27 °C, pH varied in the range of 8.0 to 8.4, dissolved oxygen in the range 8.2 to 9.1 mg L^{-1} and total alkalinity varied in the range of 222 to 228 mg L^{-1}. All the parameters were found to be within the tolerance limit of Mahseer fry.

In this experiment, ingredients used such as fish meat powder, shrimp meal, acetes meal, groundnut oil cake, rice bran and wheat flour to prepare diets were same as being used at Tata Electric Company's Fish farm, Walwhan, District Pune, Maharashtra. All the ingredients are locally available at low cost. Using these ingredients at equal proportions, three forms i.e. ball, pellet and flake were prepared and used for the experiment. Maximum weight gain (108.14%), length gain (21.15), specific growth rate (2.41) and protein efficiency ratio (0.56) of Mahseer fry reared for 30 days was observed in flake form diet. The best food conversion ratio (5.81) was observed in flake feed. Though, there was no significant difference (P>0.05) between pellet and flake but pellet and flake significantly differed from ball form diet. Maximum survival was found in flake feed (89.29%) and it was significantly differed (P<0.05) from ball and pellet forms. These results indicated that the maximum feed utilization, survival and growth of Mahseer fry was observed in flake feed followed by pelleted feed. The flake feed consists of fish meat powder, shrimp meal, acetes meal, groundnut oil cake, rice bran and wheat flour 16.66 per cent each, having 32.6 per cent protein, 9.2 per cent crude fat, total ash 14.2 per cent and moisture 5.8 per cent.

There are no reports on effect of diet forms on feed utilization, growth and survival of fry *Tor khudree.* Keshavanath et al. (1986) and Murthy & Keshavanath (1986) evaluated protein requirement of fry of *Tor khudree.* During their study they have tried diet in pelleted form. They found maximum gain in weight (15.10 g) and best FCR (2.85) in diet containing 40 per cent protein. In the present study, maximum weight gain (108.14%), best FCR (5.81) and survival (89.29%) was found in flake feed, followed by pellet feed.

REFERENCES

Chow, K.W., 1980. Quality control in fish feed manufacturing. In: *Fish Feed Technology,* Aquaculture Development and Coordination program. (FAO/ADCP/RED/8)/1 1, pp 369-385.

Kulkarni, C.V. and Ogale, S. N. 1995. *Conservation of mighty Mahseer of India.* Tata electric companies Bombay House, 7-38 pp.

Keshvanath, P., Varghese, T. J., Shetty, H. P. C., Krishnamurthy, D. & Gogoi, D.,1986.

Impact of diets with various protein sources and 17 α methyl testosterone on the growth of Mahseer, *Tor khudree* (Sykes). *Punjab Fisheries Bulletin,* 10 (2): 72-83.

Murthy, S.V. and Keshavanath, P., 1986. Protein requirement of Mahseer, *Tor khudree* (Sykes) with a note on feed utilisation. *Punjab Fisheries Bulletin,* 10 (2): 64-71.

Shirgur, G.A. and Indulkar, S.T., 1994. Fry to fingerling rearing trials of *Tor khudree* in a cement cistern and kutcha pond. In: Prakash Nautiyal (Editors), *Mahseer the game fish.* Rachna Pub, pp. D 66- D 78.

Zar, J.H. (Editor), 1974. *Biostatistical analysis.* Prentice-Hall, Inc, Enewood cliffs, N. J. U.S.A., 1-620 pp.

3

Studies on the Performance of 'Ovatide'– On Breeding of Indian Carps

M.G. Babare, M.K. Kale, M.V. Mote and S.N. Choudhary*

SUMMARY

The paper deals with the experimental studies on breeding of Indian Major Carps - Catla catla (Valenciennes), Labeo rohita (Hamilton) and Cirrhinus mrigala (Hamilton), with the help of 'Ovatide' which is being used as an alternative inducing agent for commercial seed production. This .study has been conducted for six consecutive months (April. September, 2004) in a stone - pitched breeding channels of a farm located at Osmanabad District, Maharashtra. The doses of 'Ovatide' (0.5 ml/kg of fish weight) remained same for each female species during the entire study period and males were released without any dose. The Physicochemical parameters of water during different months were estimated. The latency period and fertilization percentage varied in different months and species. The results confirmed that 'Ovatide' can be used successfully in a much more cost-effective way for induced breeding of carps, even in rural fish farms with morrumpitched breeding channels.

Keywords: Induced breeding. Indian Major Carps, Ovatide.

In recent years aquaculture has been recognized as an important strategy for Meeting the growing demands of fish protein all over the world. For the rapid development and

Department of Zoology, A.S.C. College Naldurg, Maharashtra.
* Department of Zoology, K.S.K. College Beed, Maharashtra.

expansion of aquaculture industry, the steady supply of good quality seeds of target fish species is required. The seeds collected from natural breeding grounds are not sufficient to meet the growing demands. Therefore hatchery production of seeds through controlled reproduction taking into consideration of environmental and biological parameters have been in practice (Harvey and Hoar, 1979; Peter et al. 1988; Lin et al. 1991 b). To mitigate this problem of getting good quality gland in proper time for hypophysation vis a vis finding out an effective substitute of pituitary glands, investigations have been conducted by different workers with synthetic gonadtrophin releasing hormone (GnRH) analogs 'Overprim' (Nandeesha et al. 1990 a and b; Roy, 1996) and 'Ovatide' (Reddy and Thakur, 1998). The aim of the present investigation is to study the effect of 'Ovatide' - an inducing agent of increase the spawning capacity of Indian Major Carps, mainly Catla catla (Valenciennes), Labeo rohita (Hamilton) and Cirrhinus mirgala (Hamilton) in a morrum pitched breeding channels of a fish farm of Osmanabad District, Maharashtra, India.

The experimental were conducted in Bori Fish Farm located at Village Naldurg, Dist. Osmanabad, Maharashtra, in India, where fishes were spawned in morrum pitched breeding channels and subsequently seeds were transferred to cement cistern. Males and females carps were selected based on external secondary sexual characters (Jhin). Only the females of four species. (Catla catla, Labeo rohita, Labeo calbasu and Cirrhinus mrigala) were injected intramuscularly at the shoulder region near the base of the dorsal fin (Jhingran, 1991). 'Ovatide' a synthetic GnRH analog and dipamine antagonists were injected at a does of 0.5 ml/kg prior to the spawning (Vth stage of maturity) during the entire period of study (April to September 2004). Females and males were released in the breeding channels and water flow was initially maintained for two hours and then with an interval of half an hour depending on the prevailing environmental conditions (Temperature, humidity, precipitation etc.). The breeding tests were conducted by releasing female and male in ratio of 1 : 2. Fishes were examined for breeding response after 6 hours of injection at an interval of two hours. The percentage of fertilization was calculated using samples of eggs. The extent of spawning was recorded by examining spawned females. The females were recognized as partially spawned when a large number of eggs oozed out freely by applying slight pressure on the abdomen. Similarly, females were classified as ovulated when no natural spawning had taken place and the eggs were oozing out in such cases with slight pressure on the abdomen. The quantity of eggs was estimated by collecting eggs with a cotton cloths and then placed in a container with certain capacity. The physico-chemical parameters of water were determined following standard methods (APHA-1995).

From this experimental study, the spawning response was found to be 100% for all the culturable species during July, while results showed same outcome for the month of June. No such response was recorded during September for *C. mrigala* and for *L. rohita* in the month of April. The maximum quantity of eggs and healthy larvae, were produced in the month of July followed by June, May and August for all culturable species. Percentage of fertilization did not reveal any striking difference among species and months (Table 3.1) pH was found to be more alkaline during the months of June - September when dissolved Oxygen, total hardness, alkalinity and conductivity showed higher values (Table 3.2). The latest development of induced breeding technique for enhancing commercial aqua-farming holds great promise as revealed from this investigation which has clearly shown the potential of synthetic peptide analogs to GnRH combined with dopamine antagonist. The positive response of all species especially by Labeo rohita and Labeo calbasu with a single dose of 0.5 ml/kg. To only females, is very significant for commercial seed production

However, the ecological factors which act as extrinsic stimuli (Harvey and Hoar, 1979) on the reproduction of telecast fishes played significant role as best results were observed in the present study during June-August in comparison to other three months.

The authors express their gratitude to District fishery department of Osmanabad for their help in conducting experiments. Thanks are also due to Prin. S.D. Peshve for their help in conducting experiments.

Table 3.1: Details of experimental results conducted with 'Ovatide' for induced spawning of Indian Major Carps.

Species	Months	Female		Male		Dose of Injection	Time	Spawing Response (No. & Weight)	Time of egg Laying	Quantity of Eggs (Liter & No.)	% of fertili-zation	Number of larvae	%of Larvae
		Number	Weight (Kg)	Number	Weight (Kg)								
Catla catla (Valenciennes)	Apr.	3	4.5	6	4.25	0.5ml/kg	6.30pm	1(1.5kg)	11.30pm	1.5(3000)	90	18000	60
	May	3	4.5	6	6.5			3(4.5kg)	11.30pm.	22(440000)	90	352000	80
	June	3	6	6	10			3(5kg)	11.30pm.	30(600000)	92	480000	80
	July	3	5	6	8			3(5kg)	12.15am.	36(720000)	93	633600	88
	Aug.	3	4.5	6	7			2(2.8kg)	1 am.	12(24000)	91	156000	65
	Sep.	3	5.8	6	9			1(2kg)	2am.	5(100000)	90	50000	50
Labio Rohita (Hamilton)	Apr.	3	2.5	6	3.9	0.5ml/kg	6.40pm	-	-	-	-	-	-
	May	3	3.	6	4.8			2(1-8kg)	12am	15(310000)	90	248000	80
	June	3	3.2	6	4.8			3(4kg)	12am	28(560000)	91	504000	90
	July	3	4	6	6			3(4kg)	12.30am.	36(720000)	95	676000	94
	Aug.	3	3.5	6	5.5			2(2.4kg)	1.15am.	8(160000)	93	120000	75
	Sep.	3	3.6	6	5			1(1.5kg)	4am.	4.5(90000)	85	58000	65
***Cirrhinus mrigala* (Hamilton)**	Apr.	3	2	6	2.4	0.5ml/kg	6.50pm	2(1.2kg)	12am	1.5(3 5000)	90	17500	50
	May	3	1.8	6	3			2(1.2kg)	2am.	8(160000)	90	128000	80
	June	3	2.2	6	3.4			3(2.5kg)	12.50am.	13(260000)	85	208000	80
	July	3	2.4	6	4.8			3(2.4kg)	12.15am.	20(400000)	90	333200	83.3
	Aug.	3	3	6	4.8			K1kg)	Lam.	7(140000)	90	91000	65
	Sep.	3	2.1	6	4.8			-	-	-	-	-	-

Table 3.2: Water quality parameters of Breeding channel during study period (April - September 2004).

Parameters	*April*	*May*	*June*	*July*	*August*	*September*
Temperature (°C)	30	30	29	29	29	28
pH	7.3	7.4	7.8	8.2	8	7.8
Dissolved Oxygen (mg/1)	7	7.5	8	9.5	9	9
Free CO_2 (mg/1)	10.2	12.1	9.7	9.4	8.2	8.9
Conductivity (mMho)	0.29	0.31	0.28	0.32	0.28	0.32
Chloride (mg/1)	60.02	53.01	41.2	35.8	33.1	34.08
Alkalinity (mg/1 as $CaCO_3$)	100	120	127	129	128	129
Total Hardness (mg/1)	142	140	161	160	165	182

REFERENCES

APHA *American Public Health Association.* 1995 American Water Works Association and Water Environment Federation. Standard methods for the examination of water and waste water. 19th Edition. Washington DC. 2005.

Harvey, B.J. and W.S. Hoar, 1979. *The theory and practices of induced breeding in fish. IDRC-TX le.* p. 48.

Jhingram, V.G. 1969. Review of the present status of knowledge on induced breeding of fishes and problems for future research FAO/UNDP Regional Seminar on induced Breeding of cultivated fishes. *Fri/IBCF/27,* p. 30 (mimeo).

Jhingran, V.G. 1991. *Fish and Fisheries of India* In: V.G. Jhingran (Ed.) Hindustan Publishing Corporation, India. p. 359.

Lin, H-R., Zhang, M-L., Zhang, S.M., Van der Kraak, G. and Peter, R.E. 1991b. Stimulation of pituitary gonadotropin and ovarian development by chronic administration of testosterone in female Japanese silver eel, *Anguilla Japonica. Aquaculture,* 96; 87-95.

Nandeesha, M.C., Keshavanath, P. Varghese, T.J., Shetty, H.P.C. and Gopal Rao, K. 1990a. Alternate inducing agents for carp breeding; Progress in research. In Keshavanath, P. and Radhakrishnan, K.V. (eds); *Carp seed production technology,* Special publication, Asian Fisheries Soc., Indian Branch, 2: 12-15.

Nandeesha, M.C., Gopal Rao K. Jayanna, R.N., Parkar, N.C., Varghese, T.J. Keshavanath, P. and Shetty, H.P.C. 1990b. Induced spawning of Indian Major Carps through single application of ovaprim C. In Hirano, R. and Hanyo, I. (eds); *Proc. Second Ascian, Fisheries Forum,* pp. 581-585.

Peter R.E., Lin, H.R. and Kraak, G. Van der. 1988a Induced spawning in Chinese carps. *Proceedings of the Aquaculture International Congress and Exposition, Aquaculture International Congress,* Vancouver, B.C. pp. 543-547.

Peter, R.E., Lin, H.R. and Kraak, G. Van der. 1988b. Induced ovulation and spawning of cultured freshwater fish in China; Advances in application of GnRH analogues and dopamine *antagonists Aquaculture,* 74:1-10.

Reddy, A.K. and Thakur, N.K. 1998. Ovatide - A new hormonal preparation for induced fish breeding *Fishing Chimes.,* 18(4): 28-30.

Roy, P.K. 1996. "Observation on the use of Ovaprim for induced spawning of Indian Major Carps" Advances in fish and wildlife Ecology and Biology. Published by Bansilal Kaul, 1: 65-67.

Tripathi, S.D. and Bhimachar, B.S. 1972 Hypophysation of fishes with Particular Reference to India. 80p. Jabalpur. *The Directorate of Research Services, Jawaharlal Nehru Krishi Vishva Vidyalaya.*

Tripathi S.D. and Khan, H.A. 1990. Carp seed production technology —A review. In P. Keshavanath, and K.V. Radhakrishnan, (eds).: *Carp seed Production Technology.* Special Publication, Asian Fish Soc., Indian Branch, 2: 1-11.

On the Collection of Fish Fauna of the River Subarnrekha from Domunhani to the End of Jamshedpur Township with Comments on their Conservation

M.C. Mahata[1], *S.R. Pal*[2] and *P.N. Panday*[3]

Fishes of the river Subarnrekha, one of the largest and a holi river of the Jharkhand plateau were collected from Domuhani to the end of Jamshedpur township, from June 2005 to July 2006 and were identified. The fish fauna of the area comprises of 41 species under 27 genera, 13 families and 7 orders. Of these 41 species 16 species are vulnerable and three are endangered. The vulnerable species are *Cirrhinus reba* (Ham.), *Garra gotyla* (Gray), *Garra satyendranathi* (Ganguli and Duttaj, *Labeo bata* (Ham.), *Ompok bimaculatus* (Bloch), *Heteropneustus fossilis* (Bloch), *Clarias batrachus* (Linn.), *Mystus seenghala* (Sykes), *Mystus tengra* (Ham.), *Glyptothorax tilchitta* (Ham.), *Anguilla bengalensis* (Gray & Hardw), *Anabas testudineus* (Bloch), *Mastocembelus armatus* Lac., *Macrognathus aculeatus* (Bloch), *Notopterus chitala* (Ham.) and *Glossogobius giuris* (Ham.). The endangered species includes *Garra satyendranathi* (Ganguli & Dutta), *Mystus mukherjee* (Ganguli & Dutta) and *Amphipnous cuchia* (Ham.). The most commercially important fishes of the river include 11 species which are sold @ Rs. 50-120/kg in the fish market except *Anguilla bengalensis.* These fisnes are *Labeo rohita* (Ham.). *Catla catla* (Ham.), *Notopterus sps., Chela atpar*

1. P. G. Dept. of Zoology, Jamshedpur Co-operative college, Jamshedpur-831001.
2. S.P. College Khasmahal, Jamshedpur-831002.
3. P. G. Dept. of Zoology, Ranchi University, Ranchi-834008.

(Ham.), *Labeo bata* (Ham.), *Heteropneustus fossilis* (Bloch), *Clarias batrachus* (Linn.), *Anguilla bengalensis* (Gray & Hardw), *Macrognathus aculeatus* (Bloch), *Mastacembelus armatus Lac.*, *Amphipnous cuchia* (Ham.), and *Cirrhinus mrigala (Ham.).*

As the fishes in the investigated area are captured in indiscriminate manner and unscientific methods therefore their population are fading day after day so their conservation is very essential for the protection of species diversity of fish population in the river and for the benefit of mankind also, as they are oftenly used as food by almost all classes of people of the society in the area.

Keywords : Fish fauna, Catching devices, Local names, Economic value, Genera, Vulnerable, Endangered species, Conservation

INTRODUCTION

The river Subarnarekha (Fig. 4.1) is one of the largest and a holi river of the Jharkhand plateau. It originates from South of the village Nagri near Ranchi and runs in between 23° 5′N and 87° 4′E through Ranchi and Singhbhum (East) districts of the State of Jharkhand and through a part of the districts of Mayurbhanj and Midnapur of the state of Orissa and West Bengal, covering a total distance of 414 kms. (196 miles) before meeting into the Bay of Bengal near Baliapal (Balasore, Orissa). The river houses varied type of limnofauna among which the fishes constitute a largest and important group of the limnofauna. But except the molluscan fauna of the river, no other groups of fauna of the river have been analysed up till now. Therefore, a study of fish fauna has been started by the authors after selecting different parts of the river from its origin to its meeting place in to Bay of Bengal. This paper accounts the fish fauna of the river from Domuhani (river meet, i.e. the place where river Kharkai meet with river Subarnarekha) to end of Jamshedpur township. The fish fauna for the purpose were collected from June 2005 to July 2006.

MATERIALS AND METHODS

The fish fauna for the present study were collected with help of fishermen fishing in the river from Domuhani to the end of Jamshedpur township (Fig. 4.2). The fishermen catch the fishes with help of different type fish catching devices (namely Cast net, Gill net, Box traps and Hook and line fishing devices). The fishes were also collected from the daily fish market on the road side near the old Mango bridge over river Subarnarekha for present study. The fishes collected so are preserved in 4% formalin after a incision (cut) ventrally in the belly for further study. Species have been identified following the guide lines of Day (1875-1878), Talwar and Jhingran (1991), Jayaram (1999), Menon (1999) and Mishra (1959).

OBSERVATIONS

The fishes identified from the aforesaid part of the river are listed below. The list include systematic position, their length range followed by details distribution and economic value.

ACCOUNT OF FISH FAUNA

Order - Notopteroidae Family - Notopteridae)

1. Notopterus notopterus (Pallas) 1767

1767	:	*Gymnotus notopterus* Pallas; Spec. Zool 1, (7); 40, pt. 6; Fig. 2.
Local Names		Pholi/Patola/Patara; (Hindi), Pholui (Beng.) Pulli (Oriya), Folat (Kurmali & Mundari)
Length		Average size of the fish ranges in between 20-45 cm. but it attains the length upto 60 cm.
Fin Formula		D. $_{8(1/7)}$; P. $_{17}$; V. $_{6}$; A. 100; C. 19; L. I. 225; vert.$_{30/60}$
Distribution		Found in fresh water and brackish water of India, Pakistan, Burma, Malaya Archipelago, Philippines, Siam, Thailand and Indonesia. In the study site the fish remain confined in the Tisco water supply tank made on the river Subarnarekha by constituting a cemented dike

approximately 5 m height covering the total width of the river near upper part of the Jai Prakash Narayan Setu, Mango. The tank is nearly 1000 m in length having a water height variable from 4-5 m year round. The fish is found throughout the Jharkhand plateau.

Economic Value : It is an edible fish and people consume the fish. It occurs scarcely in the river and only fond in monsoon and winter months in the fish market near old mango bridge on the river Subarnarekha. It costs rupees 60-80/kg.

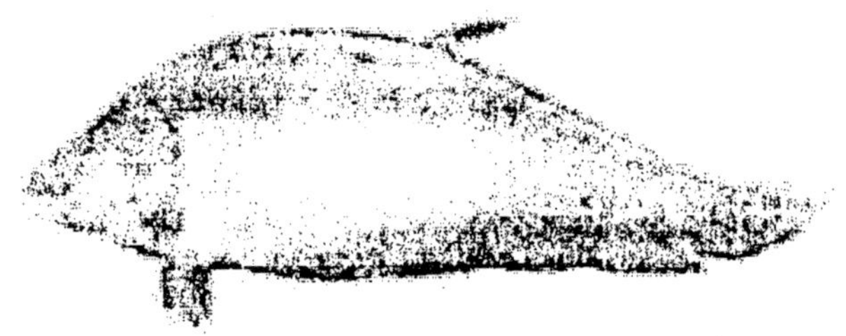

Notopterus notopterus (Pallas) 1767

---

2. *Notopterus chitala* (Ham.) 1822

1822 : *Mystus chitala* Ham., Fish Ganges; 263, 382.

Local Names Chitai (Beng.), Maya (Hindi), Pulli, Chitual (Oriya), Phalat Kurmali/Mundari).

Length Length varied from 20-25 cm, but it reaches upto a meter. Fish of this length is not found in the river.

Fin Formula $D_{9(1/8)}\ P_{16}:\ V._{6} : A._{110\text{-}118} : C._{12} : L.I._{160\text{-}180}$

Distribution Fresh water of India, Pakistan, Bangladesh, Burma, Siam, Malay Archipelago, Philippines, Thailand and Indonesia. In

Jharkhand plateau the species is found scarcely all over plateau and during monsoon to winter months it is found in the fish market near the bridge on river Subarnarekha also.

Economic Value : It is, commercially important as food. Price varies according of the size of the fish. Fish upto 250 gm cost rupees 50-80 where as fishes over 500 gm. cost rupees 80-100. Middle class people prefer the fish for consumable purpose.

Order—Cypriniformes family—Cyprinidae

3. *Oxygastor bacaila* (Ham.) 1822

1822 : *Cyprinus bacaila* Ham.; Fish Ganges, 265, 384; pl.

Local Names Chelliah, Chilwa, Chalhawa. (Hindi), Banspata (Bengali, Jellaria (Oriya), Chuhi (Kurmali/Mundari).

Length Varies from 14-17cm. and weight is varied from 15-25gm.

Fin Formula $B._{3}$: $D._{9}$; $A._{13\text{-}15}$: $L.I._{8\text{-}110}$: $Ltr_{17\text{-}19/6}$

Distribution Found in lentic and lotic water of India (Except in Malabar), Pakistan, Bangladesh. In the Jharkhand plateau, it is found in perennial tanks, lakes reservoirs and paddy field in post monsoon months. In the river the fish is abundantly found in the Tisco water supply tank on river Subarnarekha in pre-monsoon and post monsoon months and so is the case in the near by fish markets.

Economic Value : The fish is very tasty and in the market its cost is varied from Rs. 30/- to Rs. 70/- per kg. Low income group of people prefer this fish.

4. *Oxygaster clupeoides* (Bloch) 1797

1797	:	*Cypriruns clupeoides* Bloch; Ichthyogie on histoire maturelle generade at oentiuliere des poissons; pt. 2, Berlin 12, t 408.
Local Names		Shilhari (Hindi), Banspata (Beng.) Jellahri (Oriya), Chhota Chuhi (Kurmali/Mundari)
Length		The fish attains a length of 12-17cm.
Fin Formula		B.$_{3}$, D.$_{9}$, A.$_{13\cdot15}$, L.l.$_{80-3}$; Ltr.$_{12\text{-}15/6}$
Distribution		Found in the fresh water of North India, Pakistan and Bangladesh. In the Jharkhand plateau it is found through out the plateau. In investigated part of river, it is found mainly in the Tisco water supply tank.
Economic Value	:	The fish is also very tasty and provide good dish for middle class people. It costs Rs. 30-40/ kg. in the fish market. However the prices very according with availability of the fish in the market. In paucity of supply in the fish market it costs Rs. 60-100/kg. It is a larvivorous fish.

5. Oxyastor gora (Ham.) 1822

1822	:	*Cyprinus gora* Ham., Fish Ganges; 263; 384.
Local Names		Chel hul (Hindi); Gora-chella (Beng.), Humkatchari (Oriya); Bada Chuhi (Kurmali).
Length		Varies from 20-25 cm. in length.
Fin Formula		B.$_{3}$; D.$_{9\text{-}10}$; A.$_{15\text{-}16}$; LI.$_{140\text{-}160}$; Ltr.$_{18\text{-}20/18}$
Distribution		Found in the fresh water of Cutch, Jabbalpur, Mysore, the Decan, Madras and Burma. In the Jharkhand plateau the fish is found in the river Subarnarekha, Koel

	and Damoodor. In the investigated part of the river the fish is found in the Tisco Water Supply tank on the river. However its fewer occurrence was noticed below the said tank.
Economic Value :	It has a delicate flavour and is highly esteemed as food. It costs Rs.100-120/- in the fish market. It is also a larvivorous fish.

6. *Barilius bendelisis* (Ham.) 1807

1807	*Barilius bendelisis* Ham., Journey Mysore; 3; 345; pi. 32.
Local Names	Bhola (Hindi), Khoksha (Bengali), Khoksa/ Bela (Kurmali/Nagpuri/Mundari).
Length	Attains a length of 12-15 cm.
Fin Foumula	$B._{3}$; $D._{2/7}$; $P._{15}$; $V._{9}$; $A_{2\text{-}3/7\text{-}8}$; LI $_{40\text{-}43}$; L.tr.$_{7\text{-}8/5}$; $C._{18}$
Distribution :	North Eastern part of India, Pakistan, Bangladesh, Nepal and Burma. In the Jharkhand plateau the species scarcely found in the rivers. In the river it is found in the rocky bottom part.
Economic Value :	It is also a food fish and is generally consumed by poor people. It is not commercially important fish as others fishes.

7. Danio *devario* (Ham.) 1822

1822	:	*Cyprinus devario* Ham.; Fish Ganges; 341, 393, pl. 6.
Local Names		Debari (Beng.), Bonkuaso (Oriya), Patukari (Hindi), Dingil poonti (Kurmali/ Mundari).
Length		It attains a length of 4-6 cm. Sometimes it grows upto 10 cm.
Fin Formula		$B._{3}$; $D._{18-19}$; $A._{18-19}$; $LI._{41-48}$ $Ltr._{11/5}$ $C._{19.}$
Distribution		West Bengal, Bihar and Darjeeling. In Jharkhand plateau it is found all over the plateau.
Economic Value	:	Poor class of people use the fish as food. In rural areas, all classes of people consume the fish. Price varies Rs. 30-40/kg.

8. Danio rerio (Ham.) 1822

1822	:	*Cyprinus rerio* and *Chapalio* Ham., Fish Ganges, 233, 324 and 390.
Local Names		Anju (Beng.) Budu (Kurmali/Mundari).
Length		grows upto 5 cm. It avries from 1.5 cm to 5 cm.
Fin Formula		$B._{3}$, $D._{9(2/7)}$; P_{13}, $V._{8}$; $A._{15-16(2-3/12-13)}$; $C._{19}$; $L.l._{26 \cdot 28}$; $Ltr._{6}$
Distribution		Bengal, Bihar, Pakistan, Bangladesh and Burma and all over the Jharkhand plateau.
Economic Value	:	It is rarely found in the fish market. Rural people consume the species. It forms a delicious dish for them, during Winter & Summer months. They collect the fish with the help of traingular fine meshed net. Its price varies Rs. 20/- to 40/- per kg.

9. *Rasbora daniconius* (Ham.) 1822

1822 :	*Cyprinus daniconius* Ham., Fish Ganges, 327, 391, pl. 15.
Local Names	Dadhikha/Dhera/Anjani/Dankoni/ Darkina (Beng.), (Kurmali).
Length	Grows upto 14 cm in length.
Fin Formula	$B._{3}$; $D._{2/7}$; $P._{15}$; $V._{9}$; $A._{2/5}$; $L.l._{31-34}$; $L.tr._{4\frac{1}{2}/5}$; $C._{19}$
Distribution	India, Pakistan, Bangladesh, Ceylon, Burma, Malaya and Thailand.
Economic Value :	It is a larvivorous fish. Its food value is very high. It costs Rs. 40-50/kg. In the fish market it is found in the monsoon & post soon months.

10. Chela atpar (Ham.) 1822

1822 :	*Cyprinus cachius* and *atpar* Ham., Fish Ganges; 258, 259 and 283.
Local Names	Chalhwa (Hindi), Bankuaso (Oriya), Kachhi (Beng.), Chuhui (Kurmali/ Mundari).
Length	Grows upto 10 cm.
Fin formula	$D._{9}$; $A._{22-25}$; $L.I._{55-65}$
Distribution	Assam, West Bengal, Bihar, Orissa, M. P., Madras, Mysore State, Pakistan, Burma. It is found all over the Jharkhand State.
Economic Value :	It is a larvivorous fish, it is consumed by all classes of people. In the fish market the fish costs Rs. 40-80/kg. In premonsoon months it is found in the fish market.

11. *Chela laubuca* (Ham.) 1822

1822	:	*Cyprinus laubuca* Ham., Fish Ganges, 260, 384.
Local Names		Dannahrah (Hindi), Bankoe (Oriya), Layabuka/Denkena (Beng.), Choto Chuhi (Kurmali/Mundari).
Length		Grows upto 9 cm. in length.
Fin Formula		$D._{10-11}$, $A._{19-23}$, L.I. $_{34-37}$
Distribution		M. P., U.P., Bihar, Orissa, West Bengal, Jharkhand, Assam, Nepal, Bangladesh, Burma, Cylon, Sumatra, Siam and Malesya.
Economic Value	:	All classes of people consume the fish. It is rarely found in the fish market. In rural area the fish is sold at Rs. 40-60/kg in post-winter months.

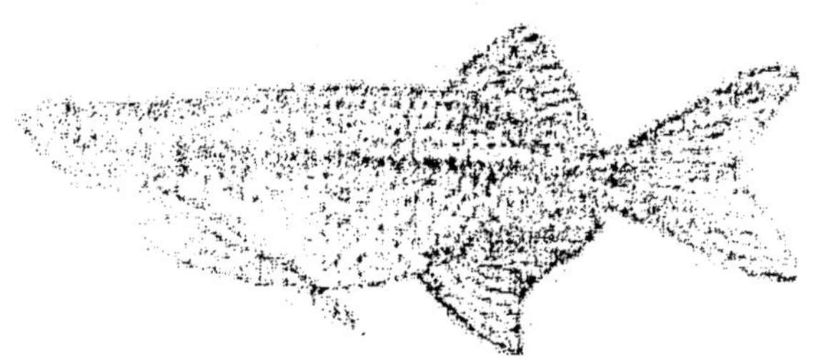

Chela laubuca (Ham.) 1822

12. *Ambypharyngodon mola* (Ham.) 1822

1822	:	*Cyprinus mola* Ham. Fish Ganges, 334, 392, pl. 38.
Local Names		Dhawni (Hindi), Morara (Oriya), Maya punti (Kurmali/Mundari).
Length		Grows upto 7.5 cm. varies from 5-7.5 cm.
Fin Formula		$D._{9(2/7)}$; $P._{15}$; $V._{9}$; $A._{7(2/5)}$; $C._{19}$; $L.I._{65-75}$; $L.tr._{12/12}$

Distribution	Throughout India, Pakistan and Burma.
Economic Value :	An important food fish of the rural areas of the Jharkhand plateau especially in the months of Mid-July-August-September and Ocotber. Costs Rs. 20-30/- per kg. Along with other smaller fishes of its categories.

13. *Catla catla* (Ham.) 1822

1822: :	*Cyprinus catla* Ham., Fish Ganges, 287, 318, 387, pl. 13.
Local Names	Katla (Hindi, Bengali, Kurmali, Mundari), Barkura/Bhakur (Oriya and U. P. & M. P.).
Length	Attain 189 cm or more in length A 25-30 cm fish commonly found in the investigated area of the river Subarnarekha.
Fin Formula	D.$_{3-4/14-16}$; P.$_{21}$, V.$_{9}$; A$_{3/5}$; L.l. $_{40-45}$; L.tr.$_{7\frac{1}{2}/9}$
Distribution	Throughout India, Pakistan, Bangladesh, Burma, Siam, and Ceylon.
Economic Value :	It is a common important food fish to all classes of people of the plateau. Its price varied Rs. 50-70/kg in the fish market and found year round in the fish market.

Catla catla (Ham.) 1822

14. *Cirrhinus mrigala* (Ham.) 1822

1822 :	*Cyprinus mrigala* Ham. Fish Ganges 279, 386, pl. 6.
Local Names	Naim/Mrigula (Hindi), Mrigal (Beng.), Mirrgah (Oriya), Midik .(Kurmali/ Mundari).
Length	It grows to 90cm in length. In the studied part of the river the maximum size of the fish varies 25-30 cm in length.
Fin formula	$B._{3}$, $D._{2\text{-}3/8\text{-}7}$' $P._{16}$, $V\text{-}_{9}$; A_{5}; $L.I._{35\text{-}38}$, $L.I._{35\text{-}38}$, $L.tr._{6½\text{-}7/8½}$; $C._{19}$
Distribution	Throughout India, Pakistan and Bangladesh. In the investigated part of the river Subarnarekha the fish is mainly found in the Tisco water supply tank year round.
Economic Value :	It is also an important food fish of all classes of people. Its price varies Rs. 70-100/kg. depending upon the size of the fish. It is found year round in the fish market nearby the river.

Cirrhinus mrigala (Ham.) 1822

15. *Cirrhinus reba* (Ham.) 1822

1822	:	*Cyprinus reba* Ham. Fish Ganges, 298, 386.
Local Names		Rewah (Hindi), Batta (Beng., Kurmali/ Mundari), Chetchua porah (Oriya).
Length		Grows upto 30 cm in length but the fish collected from the river Subarnarekha. varies 10-15/cm in length.
Fin Formula		$B._{3}, D._{2\text{-}3/9}, P._{16}\ V._{9}, A._{3/5}/L.I._{32\text{-}35}\ Ltr._{7/9}; C._{19}$
Distribution		India, Pakistan and Bangladesh.
Economic Value	:	It is good food fish for all classes of people. It is found oftenly in the near by fish market of the river and costs Rs. 50-70/kg.

16. *Garra lamta/gotyla* (Gray) 1832-33

1832-33	:	*Cyprinus gotyla* Gray, IIIrd Ind. zool. Hardnicki, 2, pl. 88, fish 3, 39.
Local Names		Ghor-pola(Beng.), Patharchata (Hindi), Pathar chata (Kurmali/Mundari).
Length		The fish attains 1 5.2 cm in length but it varies from 8-12 cm in the fish collected from the river Subarnarekha.
Fin Formula		$B._{3}, D._{2/9}, P._{16}\ V._{9}, A._{2/5}L.I._{34}\ Ltr._{4/5};$

Distribution	Assam, Darjeeling, Eastern Himalayas, Simla, E. Punjab, U.P., Western Himalayas, Bihar, Jharkhand, West Bengal, West Punjab and Pakistan.
Economic Value :	It is found scarcely in the investigated part of the river year round. It is sold with other smaller fishes (locally known as 'Chuna mach') and costs Rs. 30-50/kg. Its frequency of occurrence is more in breeding season than the other sesaon. Used as food by people of the town area.

Garra gotyla (Gray) 1832-33

17. *Garra mullya* (Sykes) 1841

1841 :	*Chondrostoma mullya* Sykes, Trans. Zool Sec. London; 2. 359, pl. 62.
Local Names	Gharpola (Beng.), Patharchata (Hindi), Choto putharchata (Kurmali/Mundari).
Length	Grows upto 12.7 cm. Its length varies from 7 cm to 1 0 cm those collected from the river.
Fin Formula	$B._{3}$, $D._{2\text{-}3/89}$, $P._{15}$ $V._{9}$, $A._{2/5}$ $L.I._{32\text{-}36}$ $Ltr._{4\text{-}4½}$; $C._{17}$
Distribution	U. P., Bihar, Jharkhand, M. P., Orissa, Chattisgarh, Kathiawar, Bombay, Poona, Mysore, Malabar, Travancore & Coachin.

Economic Value : As the species is found very scarcely, Its food value is not equal to that of the proceeding species. It is sold in the nearly fish market of the river with other smaller fishes (locally known as Chuna mach). It cost Rs. 20-40/ kg. along with other smaller fish. It is consumed by poor classes of people.

---

18. *Garra satyendranathi* (Ganguli & Dutta) 1973

1973 : *Garra satyendranathi* Ganguli & Dutta, Indian Biol; 5, 91-94.

Local Names Ghor-pola (Beng.), Patharchatta (Kurmali/ Mundari).

Length Grows upto 10.5 cm., but varies from 8.5 to 10.5.

Fin Formula $B._{3}$, $D._{11/8}$, $P._{1/11}$ $V._{1/7}$, $A._{11/5}L.I._{34}$.

Distribution River Subarnarekha of Jharkhand Plateau.

Economic Value : Similar to that of *Garra gotyla.* It is very-very scarcely occurring species.

---

19. *Labeo rohita* (Ham.) 1822

1822 : *Cyprinus rohita* Ham., Fish Ganges, 301, 388, pl. 36.

Local Names Rohu (Hindi), Ruee (Bengali, Kurmali, Mundari), Rohu (Oriya).

Length Grows more than 91.4 cm. The species in the catches of the river is found frequently. It is measured 30 cm to 40 cm.

Fin Formula $B._{3}$, $D._{3/12\text{-}13\text{-}13}$, $P._{17}$ $V._{9}$, $A._{2/5}L.I._{40\text{-}42}$ $Ltr._{61/2/9}$; $C._{19}$

Distribution Commonest major carp of the plains of India, also found in fresh waters of Burma, Pakistan and Bangladesh.

Economic value :	It is an excellent as food and of great economic importance. It is found year round in the fish market and is consumed by all classes of people. It costs Rs. 50-80/ kg, which very according with size of the fish.

Labeo rohita (Ham.) 1822

20. *Labeo bata* (Ham.) 1822

1822	:	*Cyprinus bata* Ham., Fish Ganges, 283, 386.
Local Names		Bata or Bata roee (Kurmali), Bhangra or Bata (Bengali), Dunguda-p. orah or bata (Oriya), Bata (Hindi).
Length		It grows upto 60 cm in length. It is a minor carp. The species from the river varies 25-35 cm in length.
Fin Formula		$B._{3}, D._{2-3/9-11}, P._{18}\ V._{9}, A._{2/5} L.I._{37-40}\ Ltr._{7-6-7}; C._{19}$
Distribution		Krishna and Gudawari rivers in Assam, West Bengal, Orissa, M.P., U.P., Chattisgarh, Bihar and Bangladesh.
Economic value :		It is used as food by all classes of people. It is also found year round in the fish market and cost Rs. 60-80/kg according to size and supply to the market. It is tasty one. Scarecly found in the river and is vulnerable.

Labeo bata (Ham.) 1822

21. *Puntius sarana* (Ham.) 1822

1822 :	*Cyprinus sarana* Ham., Fish Ganges, 307, 388.
Local Names	Pottah (Kurmali), Punti (Beng.), Sarana (Oriya), Potti (Mundari).
Length	Grows upto 12.9 cm. The fish from the river measures 5.5 cm onward.
Fin Formula	B.$_{3}$; D.$_{4/5}$; P.$_{15\text{-}17}$; V.$_{8\text{-}9}$; A.$_{2\text{-}3/5}$; L.I.$_{28\text{-}34}$; Ltr.$_{51/2/6/9}$; C.$_{19}$.
Distribution	India, Burma, Pakistan, Bangladesh, Nepal, Bhutan, Ceylon, Siam and China.
Economic value :	The species is sold along with other smaller fish in the near by fish market of the river. It is not oftenly occurs. It costs Rs. 30-60/ kg and is consumed by all classes people.

22. *Puntius sophore* (Ham.) 1822

1822 : *Cyprinus sophore* Ham., Fish Ganges, 310, 389, pl 19.

Local names : Katch-Karwa (Hindi), Punti (Bengali), Poonti (Kurmali), Patia Kerundi (Oriya), Karandi (Mundari).

Length : Grows upto 13 cm. The species from the river Subarnarekha varies 7-10 cm in length.

Fin Formula : $B._{3}$; $D._{3/8-9}$; $P._{17}$; $V._{9}$; $A._{3/5}$; $L.I._{23-25}$; $Ltr._{4\frac{1}{2}-5/5}$; $C._{19}$

Distribution : India, Burma, Pakistan, Bangladesh, Nepal and China.

Economic Value : As *Puntius sarana.*

Puntius sophore (Ham.) 1822

23. *Lepidocephalichthys guntia* (Ham.) 1822

1822 : *Cobitis guntea* Ham. Fish Ganges, 353, 394.

Local Names : Nakti/Gunguch (Hindi), Kendaturi/ Gupkari (Oriya), Gentu (Kurmali), Bali gada (Mundari).

Length : Grows to 8.9 cm.

Fin Formula : $D._{8(2/6)}$; $P._{8}$ $V._{7}$; $A._{7(2/5)}$; $C._{16}$; $L.I._{115}$

Distribution : Throughout India (except Mysore, South of the Krishna river and Malabar coast) and Pakistan.

Economic value	:	It has high food value to rural people. It is used as medicine and the rural people consume it in loss of apetite. The catches are found in abundant in the month of Octobar-November. It costs Rs. 100-120/ kg.

24. *Naemacheilus savana* (Ham.) 1822

1822	:	*Cabitis savana* Ham., Fish Ganges, 357, 394.
Lenght		Grows more than 7.0 cm.
Fin Formula		$B._{3}$; $D._{10\text{-}11(2\text{-}3/8)}$; $P._{10}$; $V._{7}$; $A._{7(2/5)}$; $C._{18}$
Distibution		Bengal, Bihar, Assam, Punjab and U. P.
Economic value	:	As *Lepidocephalichthyes guntea.*

DIVISION - SILURI; FAMILY - SILURIDAE

25. *Ompok bimaculatus* (Bloch) 1797

1797	:	*Silurius bimaculatus* Bloch, Syst Ichth. 11, 17, pl. 369.
Local Names		Puffta (Hindi), Pabda (Beng.), Pobtah (Oriya), Pobdah (Kurmali).
Length		Grows to 30.4 cm in length. In the river the fish varies 20-25 cm in length.
Fin formula		$B._{12\text{-}15}$, $D._{2\text{-}5}$; $P._{1/11\text{-}15}$, $V._{8\text{-}10}$, $A._{2\text{-}3/47\text{-}72}$, $C._{13\text{-}19}$
Distribution		Throughout India, Pakistan, Ceylon, Burma, Malaya, Malay-Archipelago, Siam, Yunnani, Indonesia, Thialand, Java, Sumatra, Borneo & Indo-China.
Economic Value	:	The fish is sold in the near by fish market of the river year round along with other smaller fishes. It has food value to poor people. It costs Rs. 30-40/kg in the fish market.

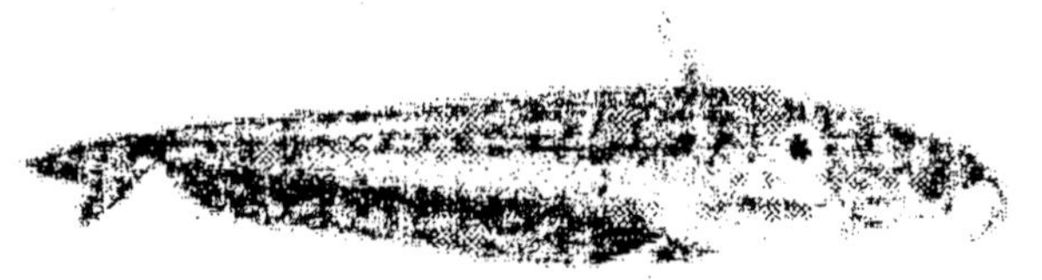

Ompok bimaculatus (Bloch) 1797

Family - Saccobranchidae

26. *Heteropneustus fossilis* (Bloch) 1785

1785	:	*Silurus fossilis* **Bloch, Ichthyol.** p.370.
Local Names		Singi (Hindi), Singhi (Beng.), Singee (Oriya), Singi (Kurmali).
Length		Grows to 30 cm or more in length. In the river it grows to 15-20cm in length.
Fin Formula		$B._{7}$; $D._{6\text{-}7}$; $P._{1/7}$; $V._{6}$; $A_{60\text{-}69}$; $C._{19}$
Distribution		India, Pakistan, Bangladesh, Burma, Ceylon, Siam, Indo-China, Nepal and Thiland.
Economic Value	:	It has high food value. It is consumed by all classes of people. It has medicinal values to rural people. It costs Rs.80-120/kg depending upon the size of the fish. It is found in all most seasons in the fish markets.

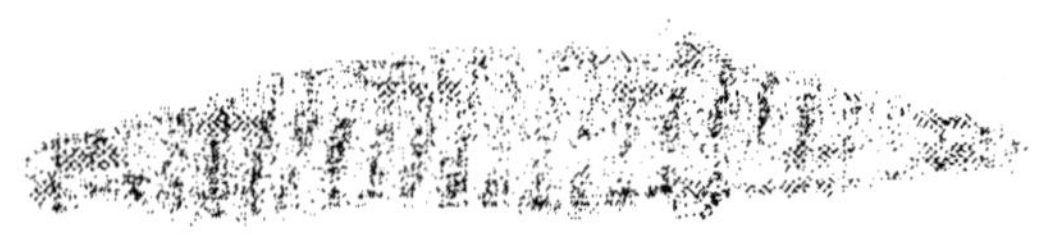

Heteropneustes fossilis (Bloch) 1785

27. *Clarius batrachus* (Linn.) 1758

1758 :	*Silurus batrachus* Linn., Syst. Nat. Led., 10, 305.
Local Names	Magri, Magur (Hindi), Mah-gur (Beng.), Magurah (Oriya), Magur (Kurmali).
Length	Attains 30.4 cm. The fish from the river grows 20-25cm in length.
Fin Formula	B.$_{9}$, D.$_{62-76}$, P.$_{1/8-9}$ V.$_{6}$, A.$_{45-58}$; C.$_{15-17}$
Distribution	Throughout India, Pakistan, Bangladesh, Burma, Ceylon, Siam and Indo-China.
Economic Value :	As *H. fossilis*. It has medicinal values and cultural values to the rural people of Jharkhand.

Family - Bagridae

28. *Mystus seenghala* (Sykes) 1839

1839 :	*Ptalystoma seenghala* Sykes, Trans zool. Soc. Lond; 2, 311, pl. 65.
Local Names	Ari (Hind), Aiir (Beng.) Alli, Addi (Oriya), Aiir (Kurmali).
Length	Attains at least 38.1 cm. The fish from the river attains the length of 25-30 cm.
Fin Formula	B.$_{12}$, D.$_{1/7/10}$; P.$_{1/91}$; V.$_{6}$; A.$_{3/8-9}$; Ltr.$_{19-2}$
Distirbution	Throughout India, Bangladesh, Pakistan, Burma and Yunan.

Economic Value : The fish is mostly found in the rainy season nearby fish market of the river. Used as food by middle and poor classes of people. It costs Rs.30-50/kg and is sold with other species of mystus. It is also a tasty fish.

29. *Mystus tengra* (Ham.) 1822

1822 : *Pimelodus tengra* Ham., Fish Ganges, 183, 377, pl. 33.

Local Names Tengra (Beng., Hindi & Kurmali) Add (Beng., Kurmali & Mundari).

Length Attains 7.5 to 10 cm.

Fin Formula $D._{1/7/9}$; $pl._{1/8}$; $V._{6(1/5)}$; $A._{11(2/9)}$; $C._{17}$

Distribution Throughout northern India, Pakistan, M. P. and Bihar.

Economic Value : As *M. seenghala.*

30. *Mystus mukherjee* (Ganguli & Dutta) 1975

1975 : *Mystus mukherjee* Ganguli & Dutta, Dr. B. S. Chauhan comm. vol., 293-298.

Local Names : Tengra (Kurmali and Mundari); Arr/Add (Beng.).

Length Attains upto 11.2 cm.

Fin Formula $D._{11/7}$, $P._{1/9}$ $V._{7}$, $A.^{11/9}$; $C._{17\text{-}18}$

Distribution River Subarnarekha in Jharkhand Plateau.

Economic Value : As *M. seenghala.*

Family - Sizonidae

31. *Giyptothorax tilchitta* (Ham.) 1822

1822 : *Pimeclodus tilchitta* Ham., Fish Ganges, 185, 378.

Local Names Telliah (Hindi), Gooacharah (Beng.) telchitti/Gulli (Kurmali/Mundari).

Length	Grows upto 15 cm is length.
Fin Formula	$D._{1/6/0}$; $P._{1/9}$; $V._{6}$; $A._{12(2/10)}$; $C._{18}$
Distribution	Delhi, U. P., Bihar, West Bengal, Bangladesh and Nepal.
Economic Value :	As the fish is very scarcely found it has less commercial values. However it is eaten by fishermen (Keut) and other rural people whenever available.

Order - Anguiliformes
Family - Anguilidae

32. ***Anguilla bengalensis*** **(Gray & Hardw) 1833-34**

1833-34 :	*Muraena bengalensis* Gray and Hardwicke, IIIrd, Ind. Zool. Hardwicke; 2, pl. 95.
Fin Formula	$D._{250-305}$; $P._{18}$; $A_{220-250}$; $C._{10-12}$
Distribution	Fresh water and seas of India Andamans, Pakistan, Burma, Malay, Archipelago, China and Bangladesh. The fish occurs in the bredding season in this part of the river.
Economic Value :	The fish is eaten by poor class people and the fisher men. According fisher men the costs of a fish is approximately Rs. 150-200 only. It is also used as medicine by the rural people of the area.

Order - Ophiocephaliformes
Family - Ophiocephalidae

33. *Ophiocephalus punctatus* (Bloch) 1793

1793 :	*Ophiocephalus punctatus* Bloch, Naturg. Ausland Fische, 7, 139. pl. 338.
Local Names	Dheridhok (Hindi), Goroi(Beng., Kurmali), Chenga (Oriya).
Length	The fish grows upto 30 cm in length.
Fin Formula	$B._{5}$; $D._{39\text{-}32}$; $P._{17}$; $V._{6}$; $A._{21\text{-}23}$; $L.I._{37\text{-}40}$; $Ltr._{4\text{-}5/9}$; $C._{12}$
Distribution	Throughout India, Pakistan , Bangladesh, Burma, Ceylon, Afghanistan, Malaya, Malay-Archipelago, Siam, Indo-China and Yunan.
Economic Value :	The fish occurs year round in the river and in the fish market. Middle and poor classes people consume the fish. The smaller size of the fish is oftenly sold with other smaller fishes (Chuna Mach) at Rs.30-40/kg. The larger fish is commercially important one and costs Rs. 60-80/kg. It is also used as medicine by the rural people and has cultural value.

34. *Channa striatus* (Bloch) 1793

1793 :	*Ophiocephalus strfetus Bloch,* Nuturg. Ausland Fische, 2, 141; pl. 359.
Local Names	Sauri (Hindi), Cheng (Beng.), Cheng/Sol (Kurmali & Mundari).
Length	Grows upto 91 cm in length.
Fin Formula	$B._{5}$; $D._{37\text{-}45}$; $P._{17}$; $V._{6}$; $A._{23\text{-}26}$; $L.I._{50\text{-}57}$; $Ltr._{41/2\text{-}8/7\text{-}10}$; $C._{13}$
Distribution	Throughout plains of India, Pakistan, Ceylon, Burma, China & Philippines.
Economic value :	As *Ophiocephalus punctatus.*

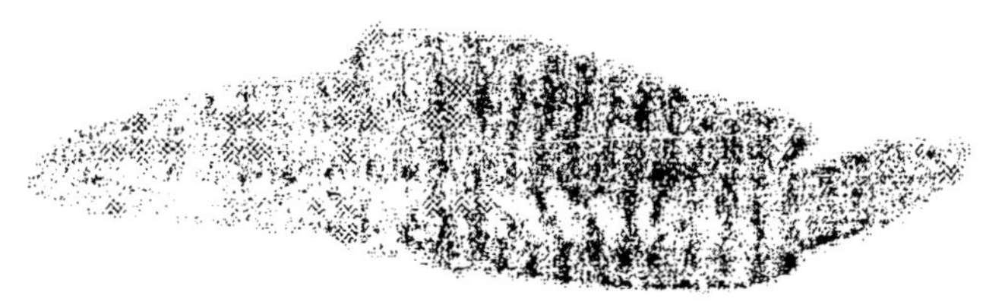

Channa striatus (Bloch) 1793

Order - Perciformcs
Family - Centropomidae

---

35. *Ambasis nama* (Ham.) 1822

1822	:	*Chanda nama* Ham., Fish Ganges, 109, 371, pl. 39.
Local Name		Chanda (Beng.), Cart Kona (Oriya), Channe (Hindi), Chanda (Kurmali and Mundari).
Length		Small fish grows upto 8 cm.
Fin Formula		$B._{6}$; $D._{7/1/13\text{-}17}$; $P._{13}$; $V._{1/6}$; $A._{3/14\text{-}17}$; $C._{17}$
Distribution		Punjab, U.P., Bihar, Darjeeling, Bengal, Orissa, M.P., Chattisgarh, Pakistan, Bangladesh, Burma & Nepal.
Economic Value	:	It costs Rs.30-40/kg in the fjsh market. The fish is sold along with other smaller fishes and is consumed my middle class of people along with other smaller fishes.

---

36. *Ambasis ranga* (Ham.) 1822

1822	:	*Chanda ranga* Ham. Fish Ganges, 113, 371, pl. 16.
Local Names		Chanda (Beng.), Loal Chandee (Oriya), Chandi (Kurmali and Mundari).
Length		Grows upto 10 cm.

Fin Formula $B._{6}$; $D._{7/1/13\text{-}15}$; $P._{11}$; $V._{1/5}$; $A._{3/14\text{-}16}$; $L.I._{60\text{-}70}$; $C._{19}$

Distribution Throughout India, Pakistan, Bangladesh, Burma, Malaya, Siam & Thailand.

Economic Value : As *Ambasis nama.*

Family - Anabantoidei

37. *Anabas testudineus* (Bloch) 1785

1792 : *Anabas testudineus* Bloch, Naturges Ausland Fische, 6, 121, pl. 322.

Local Names Coi (Beng.), Kabhai (Hindi), Koi (Oriya), Koi (Kurmali & Mundari).

Length Grows upto 14.5 cm in length.

Fin Formula $B._{6}$; $D._{17\text{-}18/8\text{-}10}$; $P._{15}$; $V._{1/5}$; $A._{9\text{-}10/9\text{-}11}$; $L.I._{28\text{-}32}$; $Ltr._{3\text{-}4/9\text{-}10}$; $C._{19}$

Distribution Bihar, Bengal, U. P., Assam, Pakistan, Bangladesh, Burma, Ceylon, Malaya, Malay-Archipelago, Siam, Indo-China, China, Philippines, Nepal, Thailand and Combodia.

Economic Value : The fish form a important and fascinated food fish to all classes of people. It cost Rs.50-70/kg in the fish market. It is a larnivorous fish.

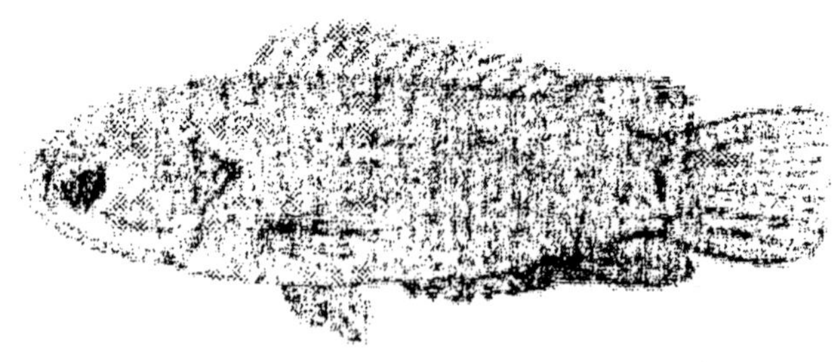

Anabas testudineus (Bloch) 1785

38. *Glossogobius giuris* (Ham.) 1822

1822	:	*Gobius giuris* Ham; Fish Ganges, 51, 366, pl. 33.
Local Names		Belay/Bhalla (Beng.), Gulah (Oriya), Bulla/ Bhula (Hindi), Bele/Khouksa (Kurmali & Mundari).
Length		Grows upto 46 cm in length. In the investigated part of the river it grows 20-25 cm only.
Fin Formula		$B._{4}$; $D._{6/1/8-9}$; $P._{20}$; $V._{6(1/5)}$; $A._{9-10(1/8-9)}$; $C._{17}$; $L.I._{30-35}$
Distribution		India, Pakistan, Bangladesh, Burma, Ceylon, E & S. of Africa, Mauritius, Malaya Malay-Archipelago, Siam, China, Japan, Philippines, Australia, New Colidonia & Thailand.
Economic Value	:	A popular food fish. It costs Rs. 50-60/kg. When available in the fish market. It is also used as medicine by rural people.

Glossogobius giuris (Ham.) 1822

Order - Mastacembeleformes
Family - Mastacembelidae

39. *Mastacembelus armatus* (Lac.) 1800

1800 : *Mastacembelus armatus* Lacepedes, Hist. Nat. poissons, 2, 226.

Local Names Boam (Beng.), Bummi (Oriya), Baam (Hindi), Bamii (Bihar), turr/ Daen (Kurmali & Mundari)

Length Attains upto 60 cm in length. The fish form the river measures 30-45 cm.

Fin Formula $B._{6}$; $D._{16\text{-}20/44\text{-}54}$; $P._{23}$; $V._{2\text{-}3/44\text{-}52}$; $A._{2\text{-}3/44\text{-}52}$; $C._{15}$

Distribution :- India, Pakistan, Bangladesh, 'Burma, Ceylon, Malaya, Siam, Nepal, Thailand, Combodia, S. China/Sumatra and Java.

Economic Value : It is highly esteemed as food. All classes of people consumed the fish. It costs Rs. 60-80/kg in the fish market. Available abundantly in the fish market after winter months.

Mastacembelus armatus (Lac.) 1800

40. *Macrognathus aculeatus* (Bloch) 1787

1787 : *Ophidium aculeatum* Bloch, Lchthyl., 5 60, pl. 159.

Local Names Gainchi (Beng.), Sand (Oriya), Patgaincha (Bihar), Thuri/Gainch/Dain (Kurmali &

	Mundari)
Length	Attains upto 38 cm in length.
Fin Formula	$B._{6}$; $D._{16-20/44-54}$; $P._{23}$; $V._{2-3/44-52}$; $A._{2-3/44-52}$; $C._{15}$
Distribution	India, Pakistan, Bangladesh, Burma, Ceylon, Malaya, Malaya Archipelago, Siam, Indo-China, China, Thailand, Java, Sumatra, Borneo & Moluccas.
Economic Value :	As *Mastocembelus armatus.*

Order - Symbrmchiformes
Family - Amphiprioidae

41. ***Amphipnous cuchia*** **(Ham.) 1822**

1822 :	*Unibranchapertura cuchia* Ham. Fish Ganges, 16 369, pl.
Local Names	Cuchia, Kunchi Beng.; Kuchia (Hindi & Kurmali), Kuchai (Mundari).
Length	Attains 60.9 cm in length.
Fin Formula	D. rudimentary, P.; V.; A.; C. absent.
Distribution	Punjab, Orissa, Darjeeling, W. Bengal, Assam of India & North Bengal.
Economic value :	The fish has food value to rural people. The fish is medicinal to rural people. It is sold Rs. 100-150/- kg when available in the market. In the studied area of the river the fish is rerely found in breeding and post breeding months.

DISCUSSION

In the present work we have recorded 41 species, under 27 genera, 13 families and 7 orders. This account allows that the water bodies from Domuhani to the end of Jamshedpur township of ihe river subarnarekha harbours by some vulnerable species like Garra *gotyla, Garra satyendranathi, Cirrhinus reba,"* *Mystus tengra, Mystus tengra Ompok bimaculatus, Clarias* ***batractws, Heteropneustus fossilis, Channa punctatus, Glossogobius***

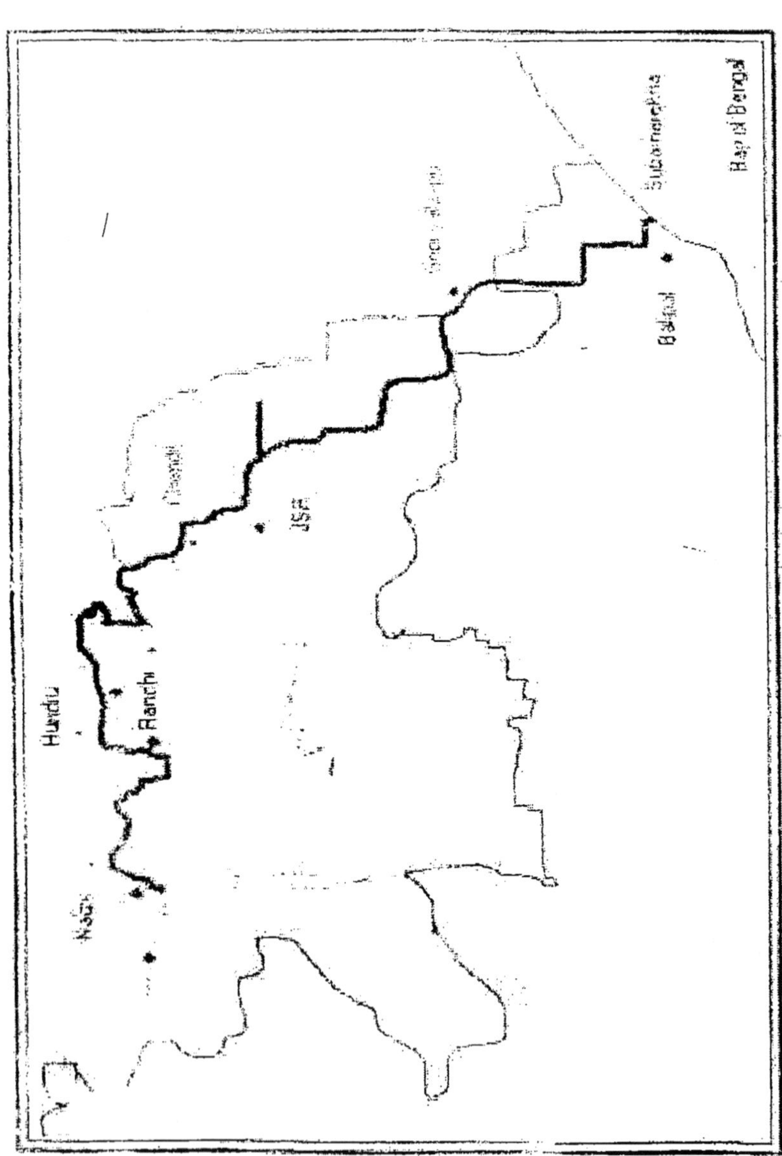

Fig. 1 : Map of the river Subamarekha

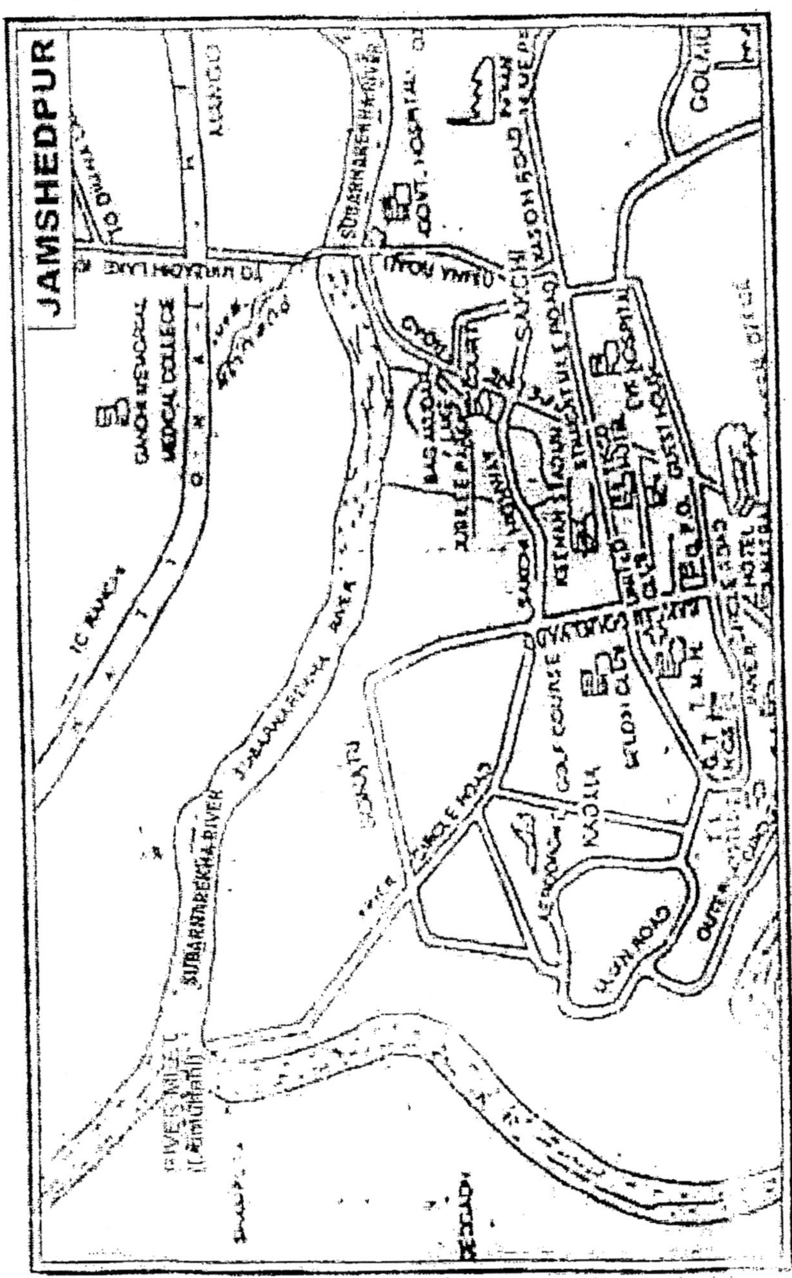
JAMSHEDPUR

giuris, Anabas testudineus, Anguila bengalensis, Notopteru emitala, Gryptelhotes' tilchitta, Mastacembe llusarmaatus and *Macrognathus aculeatus* etc. The *Garra satyendranathi* is endemic as well as endangered fish of the river. Other endangered species of the investigated part of the river are *Amphipnous cuchia* and *Mystus mukherjee.* While investigation it has been noted that high priced commercially important fishes of the over are Labeo *rohita, Catla, catla, Cirrhinus mrigata, Nototoptents sps., Chela atpar, Labeo bata, Heteropneustus fossilis, Clarias batrachus, Anguilla bengalensis, macrognathus aculeatus, Mastacembelus armatus* and *Amphipnous cuchia.*

The fishes of Chotanagpur plateau was studied previously by K.C. Bose, M.C. Mahata, P.N. Panday and A.C. Goral (1976-77) and have reported 81 species belonging to 47 genera under 2f families and 8 orders. Bose *et al* (1974-75) have reported 41 species belonging to 28 genera under 16 families and 8 orders from Jarnshedpur. The species which were not reported by them includes *Notopterus chitala, Labeo rohita, Labeo bata, oxygaster clupeoides, Oxygastor gora, Danio rerio, Chela atpar, Amblygopharygodon mola,* Garra *gotyla, Garra mullya, G Satyendranathi, Maystus mukherjee.*

However the fishes which were not found durng the investigation in the investigated part of the river but are reported by Bose *et al.* (1974-75) include *Botia dario. Oxygastor phulo, Baritius bola, Puntius conchonius, P. ticto, Bagarius bagarius, Gumbusia affinis, Channa gachua, Channa maralius, Colisa faciata* and *Mastacembelus panaculus.*

Bose and Sinha (1974-75) reported 43 species of fishes under 27 genera, 14 families and 7 order from Chiabasa. Many species which are found in fresh water systems of Chiabasa are not found in the river Subarnarekha (from Domuhani to the end of Jamshedpur township) include *Labeo calbasu, L. boga, Puntius ticto, P. amphibia, P. stigma, P. pinnauratus, Rohtee belongeri, Botia dero, Wallago attu, Mystus cavasius Channa gachua, C. marulius, Ambassis nama, Ambassis bacuilis* and *Anguila bengalensis,* Gorai and Mahata (1974-75) reported 33 species of fishes from Purulia an adjacent area to Singhbhum (Jharkhand). Some species those are found in Purulia area are also found in the studied area of the river Subarnarekha. They belong undar 22 genera, 12 families and 5 orders. The species these one found in purulia, and are not found in the river are *Labeo calbasut Wallago attu, Silonia silondia, Alia coila, Ctupisoma garuat Mystus* wtefus, *Channa gachua, C. striatus* and *Mastacembelus panaculus;*

Further a comprehensive account of the ichthyofauna of Cauvery River system was given by Jayaram *et al.* (1982), listed the presence of 93 species from Karnataka. Recently Chandrasekhariah *et.al.* (2000) have given elaborate vision of distributional pattern and status of the Ichthyofauna in different drainage in Karnataka, viz. Cauvery, Krishna, Godawari and West flowing river in the State. They reported 97 species from Cauvery river including 42 species not reported by Jayaram. The ichthyofaunal composition of the Cauvery system draining Karnataka are more or less similar to the fishes of present investigated sites of the river Subarnarekha. These fauna also shows great similarity in occurrence with fish from Bangalore and Kolar districts, Karnataka (Krishna *et al.* 2004) and Kolleru lake, Andhra Pradesh (Burman 2004).

Conservation : For the conservation of fish resources of the river Subarnarekha along with its vulnerable, endangered and rare fish species following measures be undertaken.

1. Indiscriminate fishing should be stopped.
2. Restrictions on the use of certain nets during a particular time of the year from April to June.
3. Restriction on the capture and sale of legal sizes of fry and fingerlings of some fishes such as *Garra gotyla, Nemachilus sps., Labeo buta, Heteropneustus fossilis, Clarious batachus* and *Anabus tistudineus* and major and minor camps.
4. Direct discharge of industrial effluents from the industries of Jamshedpur town should be stopped.
5. Mass awareness campaign should be organized to educate about the significance of the conservation of the fishes in the areas, specially to fisherman.
6. Pesticides and agricultural fertilizers along the water courses of the river should be used very carefully.

REFERENCES

Berg. L S. 1940: Classification of fishes both recent and fossi; Treb. Inst. Zoo!. Acad, Sci. U.S.S.R. Levingrand; 5; 87-517.

Bose K.C., M.C. Mahata, 1976-77 : P. N. Panday and A. C. Gorai Paikilothermons vertebrates of Chotonagpur plateau, Bihar; Res. Jour. R. U.; 12 and 13; 301-316.

Bose, K.C., M. Firoz and B. Chakraborty 1974-75 : Fishes of Jamshedpur; Res. Jour. R. U.; 10 and 11; 296-3047.

Bose, K.C. and A. K. Sinha 1974-75 : Studies or Ichthyofauna of Chaibasa; IBID; 10 and 11; 296-304.

Burman, R. P. 2004 : Threatened and endemic fishes of Tripura with comments on their conservation; Rce. Zool. Surv. India; 103 (Part 1-2); 75-81.

Burman, R. P. 2004 : The fishes of the Kolleru Lake, Andhra Pradesh, India with comments on their conservation. Rec. Zool. Surv. India; 103 (Part-2) 83-89.

Chandrasekhariah, H. N. 2000 Status of Fish fauna in Karnataka; pp.-98-136, in Ponnaih, A. G and *et al.* Gopla Krishnan, A. (Eds.) Endemic fish diversity of Western Ghats, NBFGR - NAPT Publication; 1-347 pp. National Bureau of Fish Genetic resources, Lucknow, U. P., India.

Day, F. 1875-1878 : The fishes of India : being a natural history of the fishes known to inhabit the seas and freshwater of India, Burma and Ceylon, Taylor & Francis, London, 778.

Day, F. 1889 : The fauna of British India including Ceylon and Burma.

Fish I & K, Talyer and Francis, London.

Gorai, A. C. & M. C. Mahata 1974-75 : Ichthyofauna of Purulia; Res. Jour. R. U. ; 10 and 11; 314-320.

Jayaram, K. C. 1999 : /The fresh water fishes of Indian region; Narendra publishing

House, Delhi, XXVII + 551.

Jayaram K. C. 1981 : The fresh water fishes of India, Pakistan, Bangladesh, Burma and Sri Lanka, A hand book, Zool. Surv. of India, Govt. of India, XXII+ 475.

Krishnan, S 2004: On a collection of fish from Bangalore and Kolar districts, Karnataka; Rec. Zool. Surv. India; 103(1-2); 143-155.

Mishra, K. C. 1959 : An aid to the Identification of the common commercial fishes of India and Pakistan; Rec. ind. Mus.; 57 + XXVI + 318.

Menon, A. G. K. 1999 : Checklist of the fresh water fishes of India; Rec. Zool. Surv. India, occ. paper No. 175; 366, Zool. Surv. of India, Govt. of India.

Sen, Nibedita 2003 : Fish fauna of North east india with special reference to endemic and threatened species; Rec. Zool. Surv. India; 101 (Part 3-4); 81-99.

5

Qualitative Studies on Community Composition of Aquatic Invertebrate Fauna of Two Ponds of Nagpur (MS) with Special Reference to Fish Production Potentiality

Nupur E. Paul, B. Mahto** and P.N. Pandey***

SUMMARY

Fish production from marine resources has reached a plateau in India and only marginal increase may be expected in near future but inland water resources have immense potential for fish production. Fore only 16% of the available freshwater area is being utilized for fish production. The ponds used for fish production in unmanaged way and ponds not used for fish production, outnumber the managed ponds and hence have huge scope to enhance fish production. With a view to examine fish production potentiality, qualitative analysis of aquatic invertebrate fauna of two ponds of Nagpur (Maharashtra)—one being Unmanaged and other Unutilized pond, has been carried out. Replicate random samples of aquatic invertebrates were collected at monthly interval from the two ponds for a period from April '05 to March '06. Benthos were sampled with the help of Ekman's dredge, plankton with plankton net and larger insects with hand pond-net. The processed samples were preserved in 70% alcohol or 5% formaldehyde. The samples were examined under compound microscope and identified to the lowest possible taxonomic level.

* Inland Fisheries Department, Hislop College, Nagpur (Maharashtra).

** Department of Zoology, Ranch! University, Ranchi (Jharkh.inu).

Altogether 61 taxa of Aquatic invertebrates were recorded from the proposed study area of which 49 taxa belonged to macroinvertebrates (both benthic and free swimming), and 12 taxa belonged to zooplankton. Of the 49 taxa of macro-invertebrates, 13 taxa belonged to Gastropoda, 27 taxa to Insecta and 9 taxa to Gastopoda. Of the 12 taxa of zooplankton, 6 taxa belonged to Rotifera and Crustacea each. The result revealed dominance of Oligochaeta and Chironomidae among macroinvertebrate fauna, while dominance of none is reflected in zooplanktonic fauna. Further, zooplankton, Oligochaeta, Chironomidae, larvae or nymphs of Ephemeroptera, Megaloptera and Plecoptera appeared to be principal food for fish fry and adult fish. Whereas, the adults of Coleoptera, Hemiptera and odonatan nymphs appeared to be enemies of fish fry and small fish. They completed with fish for food also.

Key words: Aquatic invertebrates, Macrophyte, benthos, plankton macroinvertebrates, unmanaged pond, unutilized pond.

INTRODUCTION

India has abundant resource for fish production. Fish production from marine resources has reached a plateau in India and only a marginal increase may be expected in near future (Krishnan et. al., 2000). Thus, we have to rely on the in land water resources which have immense potential for fish production. Only about 16% of the available freshwater area is being utilized for fish production, while 84% is still left unutilized. Of course, the fish production from inland fisheries has been increasing at the rate of 5.24% per year since 1950-51 but till date full capacity utilization has not yet been achieved. Besides managed ponds, there are unmanaged ponds which are used for fish production adopting traditional technology. In such unmanaged ponds the fish fry are traditionally introduced by the fish-farmers without considering the available nutrients and the- predators inhabiting the ponds, so the fish fry introduced are left at the mercy of predators. There are still another category of waterbodies which are totally unutilized with a view to pisciculture. From such ponds, the naturally occurring fish are harvested by local people. The number of such unmanaged and unutilized ponds outnumber the managed ponds are have huge potential with regard to enhance fish production. Hence, with a view to examine the fish production potentiality, qualitative analysis of aquatic invertebrate fauna of a representative unmanaged pond and an unutilized pond of Nagpur (Maharashtra) has been carried out. Invertebrate fauna of a pond exert its impact on fish production in many ways. It serves as principal nutritive fauna for fish, some of the invertebrates compete with fish for food while others are predators and enemies to small fish, fry and fingerlings.

MATERIALS AND METHODS

Study Area

Maharashtra having an area of 307,690 Square Kilometres, and extending from 15°35' to 22°2' North latitudes, and from 17°45′ to 80°45' East longitudes exhibit diversified agroclimatic conditions. Its coastal zone has almost constant temperature throughout the year whereas, the hilly Sahyadri Zone is exposed to heavy rain fall with cool summer. Its Eastern Zone has hot, dry, sub-humid climate with assured average annual rainfall of 365 mm. The proposed study is located in Nagpur district, which constitute a part of Eastern Zone. Nagpur is situated at an elevation of 900 ft. to 1000 ft. from the sea-level. Except for summer months from March to June, which are extremely hot, the weather all round is moderate. Nagpur district has few lakes and several ponds. Based on fish-production practices the ponds of Nagpur are divided into two categories: first, unmanaged ponds, where fish production is done by fish-farmers adopting traditional methods and second, unutilized ponds, which are not utilized for fish culture, rather naturally occurring fishes are harvested by local people.

For the proposed study, two representative perennial ponds, namely Unmanaged Pond and Unutilized Pond were selected in the study area and sampling of benthos, plankton and aquatic insects were carried out at monthly interval from each of the two ponds for a period from April ′05 to March ′06.

Sampling of Benthos

Random replicate bottom samples were collected from each of the two ponds with the help of Ekman's Dredge (area of cross-section 6" x 6"). The samples were sieved (B.S. No. 60), washed and benthic macroinvertebrates were hand picked to separate from debris and then preserved in 70% alcohol for qualitative analysis.

Sampling of Aquatic Insects

Samples of aquatic insects were collected with the help of hand pond-net of mesh size 6 for adult and larger insects and mesh size 24 for nymphs and smaller insects. The insects were separated from debris and preserved in 5% formaldehyde.

Sampling of Plankton

Replicate planktons samples were collected on random basis with help of plankton net mesh-size being 40. For each sample, 50 litres of pond water were filtered through the plankton net and the volume of concentrated plankton sample was adjusted to 50 nil. The concentrated samples were preserved in 5% formaldehyde.

Identification

The specimen of aquatic invertebrates were identified up to the lowest possible taxonomic level with help of taxonomic key by Needham and Needham (1972), Pennak (1953), Mason Jr. (1973), Stimpson et al. (1982) and Hiltunen and Klemm (1980).

OBSERVATION

Present investigation revealed that the aquatic invertebrate fauna of two ponds of Nagpur were qualitatively the same and comprised of 61 taxa. Of the 61 taxa, 49 taxa constituted macroinvertebrate fauna living in or on bottom of the ponds, swimming on water surface or water column, or associated with aquatic macrophytes. Whereas, remaining 12 taxa constituted fauna floating in water column, as zooplanktonic community. List of macroinvertebrate taxa and zooplanktonic taxa recorded from the proposed sites along with their relative abundance is illustrated in Tables 5.1, 5.2, 5.3 and 5.4.

Macroinvertebrate Fauna

The macroinvertebrate fauna was found to be comprised of Oligochaeta, Insecta and Mollusca.

Oligochaeta

Oligochaetes were represented by 13 taxa belonging to three families namely, Tubificidae (9 taxa), Naididae (3 taxa) and Aeolosomatidae (1 taxa) and are shown in Table 5.1. It was observed that *Aulodrilus* sp. of family Tubificidae, *Dero obtusa* of family Naididae and *Aeolosoma* sp. of family Aeolosomatidae were abundant; *Limnodrilus angustipenis, Glossiphonia complanata* and *Lumbriculus* sp. of family Tubificidae were common while *L. claparidianus* and *Stylaria lacustris* of family Tubificidae, *Dero pectinata* and *Chaetogaster* sp. of family Naididae were rare in both the ponds. Besides this, it was also noticed that *T. tubifix, Branchiura sowerbyi* and *Limnodrilus udekenianus* of family Tubificidae were common in unmanaged pond but were abundant at unutilised pond.

Insecta

Insects were observed to be represented by 27 taxa belonging to eight orders and are shown in Table 5.2. Of the 27 taxa, 11 taxa belonged to single order Diptera while the remaining 16 taxa were distributed among seven orders, namely Hemiptera (2 taxa), Ephemeroptera (3 taxa), Megaloptera (1 taxa), Plecoptera (1 taxon), Trichoptera (4 taxa), Odonata (1 taxon) and Coleoptera (4 taxa). Diptera was not only the most diversified insectan group but also one of the most dominant components of benthic community and as such deserves separate treatment from other insects.

Among Dipterans, family Chironomidae was found to be most diversified group having six taxa belonging to three sub families, e.g. Chironominae (3 taxa), Orthocladinae (1 taxon) and Tanypodinae (2 taxa); while the remaining 5 dipteran taxa belonged to 5 families, namely, Ceratopogonidae, Tripulidae, Tabanidae, Stratiomyidae and Liriopeidae, each having one taxon. Among the six chironomid taxa, *Chironomus* sp. was most abundant, *Polypedilum* sp. and *Keifferulus* sp. were common, and *Orthocladius* sp. and *Procladius* sp. were rare at both the ponds. Further, *Tanypus* sp. was common at unmanaged pond but was abundant at unutilised pond. Among other dipterans, Ceratopogonidae was common but representatives of Tripulidae, Tabanidae, Stratiomyidae and Liriopeidae were very rare in both the ponds. These taxa were recorded only during winter months i.e. December to February.

Remaining 16 taxa of insects belonging to seven orders showed their intermittent presence in poor density throughout the year. Among them *Caenis* sp., *Ephemerella* sp. and *Cloeon* sp. of order Ephemeroptera, and *Nemura* sp. of order Plecoptera were common at both the ponds but were more frequent at

Table 5.1 : Oligochaete taxa recorded from two ponds of Nagpur (Maharashtra) and their relative abundance.

Index : (+++) = abundant, (++) = common, (+) = rare

	Unmanaged pond	*Unutilised pond*
Sub-class Oligochaeta		
Family : Tubificidae		
Tubifex tubifex (Miill)	++	+++
Lumbriculus sp.	++	++
Branchiura sowerbyi (Bedd.)	++	+++
Limnodrilus angustipenis (Brink & *Cook)*	++	++
Limnodrilus claparedianus (Rat)	+	+
Limnodrilus udekemianus (Clap)	++	+++
Aulodrilus sp.	+++	+++
Stylaria lacustris (L.)	+	+
Glossiphonia complanata (L.)	++	++
Family Naididae		
Dero pectinata (Aiyer)	+	+
Dero obtusa (Aiyer)	+++	+++
Chaetogaster sp.	+	+
Family: Aeolosomatidae		
Aeolosoma sp.	+++	+++

unutilised pond. *Sialis* sp. of order Megaloptera, *Rhyacophila* sp. of order Trichoptera, dragonfly numphs of order Odonata, *Hydrobius* sp. and *Gyrinus* sp. of order Coleoptera and *Notonecta* sp. and *Nepa* sp. of order Hemiptera were rare at unmanaged pond but common at unutilised pond. It was observed that *Triaenodes* sp., *Leptoceres* sp. and *Limnophilus* sp. of order Trichoptera; and *Berosus* sp. and *Dytiscus* sp. of order Coleoptera were very rare at unmanaged pond and rare at unutilised pond.

Gastropoda

Gastropods were found to be represented by eight taxa belonging to five families, namely Thiaridae (2 taxa) and Viviparidae (1 taxon) of order Mesogastropoda,; and family Lymnaeidae (2 taxa), Planorbidae (2 taxa) and Amnicolidae (2 taxa) of order Basomatophora (Table 5.3). All eight taxa of gastropods were common in both the ponds except *Indoplanorbis exustus* and *Gobbia orscula* which were relatively more common at unutilised pond.

It was also noticed that almost all taxa of macroinvertebrates recorded in this study constituted benthic community of the pond except adult coleopterans (e.g. *Gyrinus* sp. and *Hydrobius* sp.), adult hemipterans (e.g. *Notonecta* sp. and *Nepa* sp.) and nymphal odonatans. These insects were found swimming on water surface or in water column. Besides them, trichopterans like *Triaenodes* sp., *Leptocercus* sp. *and Limnophilus* sp. being case-dwelling organisms, were found associated with aquatic macrophytes also, in addition to bottom of the ponds. Similarly, larvae of Ephemeroptera, such as *Caenis* sp., *Ephemerella* sp. and *Cloeon* sp. and that of Plecoptera, such as *Nomura* sp. were found swimming around macrophytes as well as living on the bottom of the pond.

Zooplanktons

Altogether 12 taxa of zooplanktons were recorded from the proposed study area. Out of them, 6 taxa belonged to Rotifera and 6 taxa to Crustacea (Table 5.4). *Brachionus* sp. and *Keratella* sp. among rotifers; and *Cyclops* sp. and *Daphnia* sp. among crustaceans were found to be abundant, while *Anuraea* sp., *Asplanchna* sp., *Synchaeta* sp. and *Monostyla* sp. among rotifers; and *Mesocyclops* sp., *Diaptomus* sp., *Ceriodaphnia* sp. and Bosmina sp. among crustaceans were common at both the ponds.

DISCUSSION

In the present investigation macroinvertebrates were found to be represented by 49 taxa which is almost equal to the number of taxa recorded from other lentic systems (Jhingran, 1982; Pandey, 1985) but was considerably low than the number of taxa recorded from lotic systems (Tsui *et al*, 1978; Rosillon, 1985; Goswami et al., 2002). The higher number of taxa in lotic

Table 5.2 : Insect taxa recorded from two ponds of Nagpur (Maharashtra) and their relative abundance.

Index : (+++) = abundant, (++) = common, (++*) = more common, (+) = rare, (+*) = very rare

Taxa recorded	*Unmanaged pond*	*Unutilised pond*
Class - Insecta		
Order - Diptera		
Family - Chironomidae		
Sub-family - Chironominae		
Chironomus sp.	+++	+++
Polypedilum sp.	++	++
Kiefferulus sp.	++	++
Sub-family : Orthocladiinae		
Orthocladius sp.	+	+
Sub-family : Tanypodinae		
Tanypus sp.	++	+++
Procladius sp.	+	+
Family - Ceratopogonidae	+	++
Family - Tripulidae		
Elliptera sp.	+*	+*
Family - Tabanidae		
Tabanus sp.	+*	+*
Family - Strotiomyidae	+*	+*
Family - Liriopeidae	+*	+*
Order - Hemiptera		
Notonecta sp.	+	++
Nepa sp.	+	++
Order - Ephemeroptera		
Caenis sp.	++	++*
Ephemerella sp.	++	++*
Cloeon sp.	++	++*

Taxa recorded	*Unmanaged pond*	*Unutilised pond*
Order - Megaloptera		
Sialis sp.	+	++
Order - Plecoptera		
Nemura sp.	++	++*
Order - Tricoptera		
Triaenodes sp.	+*	+
Leptocercus sp.	+*	+
Limnophilus sp.	+*	+
Rhycophila sp.	+	++
Order - Odonata		
Dragonfly nymphs	+	++
Order - Coleoptera		
Berosus sp.	+*	+
Dytiscus sp.	+*	+
Gyrinus sp.	+	++
Hydrobius sp.	+	++

systems may be a consequence of their permanence, which provides ample opportunity to the inhabitants for evolutionary changes and diversity.

The important feature arising from this study is that certain taxa were more frequent at unutilised pond as compared to the unmanaged pond. *Tubifex tubifex, Branchiura sawerbyi, Limnodrilus udekemianus* among oligochaetes and *Tanypus* sp. among chironomids were common at unmanaged pond but were abundant at unutilised pond. *Notonecta* sp. and *Nepa* sp. among hemipterans, *Sialis* sp. among megalopteran, *Rhyacohila* sp. among trichopteran dragonfly nymphs among odonatan, *Gyrinus* sp. and *Hydrobius* sp. among coleopteran were rare at unmanaged pond but common at unutilised pond. Similarly, *Caenis* sp, *Ephemerella* sp. and *Cloeon* sp. among ephemeropterans, *Nemura* sp. among plecopterans, *Indoplanorbis exustus* and *Gobbia orscula* among molluscs were common at unmanaged pond but

Table 5.3 : Molluscan taxa recorded from two ponds of Nagpur (Maharashtra) and their relative abundance.

Index : (++) = common, (++*) = more common

Taxa recorded	*Unmanaged pond*	*Unutilised pond*
Phylum - Mollusca		
Order - Mesogastropoda		
Family — Thiaridae		
Thiara tuberculata (Muller)	++	++
Thiara scabra (Miller)	++	++
Family - Viviparidae		
Vivipara bengalensis (Gauld)	++	++
Order - Basomatophora		
Family - Lymnaeidae		
Lymnaea acuminat&(Gray]	++	++
Lymnaea ovata (Drop)	++	++
Family - Planorbidae		
Indoplanorbis exustus (Deshyayi)	++	++*
Gyraulus convexiusculus (Hutton)	++	++
Family - Amnicolidae		
Diagoniostoma cerameopdma (Benson)	++	++*
Gobbia orscula (Nevill)	++	++*

relatively more common at unutilised pond. Likewise, *Triaenodes* sp., *Leptocercus* sp. and *Limnophilus* sp. among trichopterans, *Berosus* sp. and *Dytiscus* sp. among coleopterans were very rare at unmanaged pond but simply rare at unutilised pond. Higher frequency of these taxa at unutilised pond may be attributed to several reasons, such as habitat preference, mode of life,

Table 5.4 : Zooplankton taxa recorded from two ponds of Nagpur (Maharashtra) and their relative abundance.

Index : (+++), abundant, (++) = common

Taxa recorded	*Unmanaged pond*	*Unutilised pond*
Rotifera:		
Brachionus sp.	+++	+++
Anuraea sp.	++	++
Asplanchna sp.	++	++
Synchaeta sp.	++	++
Keratela sp.	+++	+++
Monostyla sp.	++	++
Crustacea:		
Cyclops sp.	+++	+++
Mesocyclops sp.	++	++
Diaptomus sp.	++	++*
Daphnia sp.	+++	+++
Ceriodaphnia sp.	++	++*
Bosmina sp.	++	++

feeding habit, availability of food and degree of competition. It was observed that the littoral zone of unutilised pond was richly vegetated by aquatic macrophytes, such as *Ludwigia perenis, Ludwigia octovalvis, Potamogeton crispus, Digitaria* sp. and *Cyperus* sp. Same composition of quatic macrophytes was found in unmanaged pond but was sparsely vegetated. Higher frequency of *Tubifex tubifex, Branchiura sowerbyi, Caenis* sp., *Ephemerella* sp., *Cloeon* sp., *Indoplanorbis exustus* and *Gobbia orscula* at unutilised pond might be attributed to the presence of rich aquatic macrophytes. Macroinvertebrates variously use macrophytes as food or feeding ground, as substrate for attachment, swarming and egg laying. All these activities of macroinvertebrates are associated with the morphological features, physiological state and surface area of aquatic macrophytes suitable for animal colonisation (Choudhary et. al., 2002). The observed higher frequency of case-spinning trichopterans, viz. *Triaenodes* sp., *Leptocercus* sp. and *Limnophilus* sp. at unutilised pond appears to be consequence of rich aquatic macrophyte cover, which provided them case-building material and additional habitat. Several studies have revealed positive correlation between macroinvertebrate community and aquatic macrophytes (Soszka, 1975; Voigts, 1976; Rai and Sharma, 1991). Further the observed higher frequency of *Limnodrilus udekemianus* and *Sialis* sp. at unutilised pond might be due to its organically rich substrate. The death and decay of macrophytes is expected to enrich the pond bottom with organic matter. Furthermore, the higher frequency of *Tanypus* sp., *Notonecta* sp., *Nepa* sp., *Berosus* sp., *Rhyacophila* sp. and dragonfly numphs at unutilised pond may be related to their predatory food habits, availability of food organisms and lower degree of competition. It was observed that the unutilised pond was never used for fish-farming. Consequently the insect predators did not have to compete with larger fish for food.

Next noteworthy feature arising from the study is that the most dominant groups among macroinvertebrates were Oligochaeta and Chironomidae. Among oligochaetes, *Aulodritus* sp., *Dero obtusa* and *Aeolosoma* sp. were dominant taxa; whereas, among chironomids, *Chironomus* sp. was the only, dominant taxon. Further, the zooplanktonic community was found to be dominated by *Brachionus* sp. and *Keratella* sp. of Rotifera and *Cyclops* sp. and *Daphnia* sp. of Crustacea. In the previous studies somewhat similar results were obtained by Benzie (1984) and Swink and Novotny (1985). It was observed that zooplanktons, oligochaetes, chironomids, larval ephemeropteran, mega lopteran and plecopterans appeared to be principal food for fish fry and adult fish. Whereas, the adults coleopterans, hemipterans and odonatan nymphs appeared to be enemies of fish fry and small fish. They competed with fish for food also.

REFERENCES

Benzie, J.A.H. 1984. The colonization mechanism of stream benthos in a tropical river, Menik Ganga (Sri Lanka). *Hydrobiologia.* III: 171-179.

Choudhary, S.K., Sharma, G.S. and Pandey, P.N. 2002. Studies on the aquatic macroinvertebrate community structure with different weed communities of three tropical freshwater ponds of Ranchi (Jharkhand). *In:* Pandey, B.N., Choudhary, R.K. and Singh, B.K. (eds.) Biodiversity, conservation, environmental pollution and ecology, Vol. I. APH Publishing Corporation, New Delhi, pp. 75-81.

Goswami, N.D., Singh, L.B., Goswami, A. and Pandey, P.N. 2002. Qualitative analysis of benthic macroinvertebrate of Subarnrekha river at Ranchi (Jharkhand State). *In:* Pandey, B.N., Choudhary, R.K. and Singh, B.K. (eds.) Biodiversity, conservation, environmental pollution and ecology, Vol. I. APH Publishing Corporation, New Delhi, pp. 75-81.

Hiltunen, J.K. and Kelmn, DJ. 1980. A guide to Naididae (Annelida: Clitellatea : Oligochaeta) of North America. Environmental monitoring and support laboratory, Office of Research and Development, U.S. Environmental Protection Agency, Cincinnaty, Ohio, pp. 1-48.

Jhingran, V.G. 1982. Fish and fisheries of India. Hindustian Pub. Corp. (India), pp. 954.

Krishnan, M., Pratap, S. B., Ponnusamy, K., Jumaran, M.K. and Singh, H. 2000. Agriculture development in India : problems and prospects. Working proceedings held at national centre for agricultural economics and policy research, New Delhi.

Mason, Jr. W.T. 1973. An introduction to identification of chironomid larvae (ed.) U.S. Environmental Protection Agency, Cincinnaty, Ohio, pp. 1-90.

Needham, J.C. and Needham, P.R. 1972. A guide to the study of freshwater biology. Holden-Day Inc. San Francisco, pp. 108.

Pennak, R.W. 1953. Freshwater invertebrates of the United States. The Ronald Press, New York. pp. 769.

Pandey, P.N. 1985. Studies on certain insects found in composite fish culture ponds and their role in fish culture. Unpublished Ph.D. Thesis, R.U. Ranchi.

Rai, D.N. and Sharma, U.P. 1991. Correlation between macrophytic biomass and macroinvertebrate community structure in wet lands of North Binar. *Internatl. J. Ecol. Environ. Sc. (17)*: 27-36.

Rosillon, D. 1985. Seasonal variation in the benthos of chalk treat stream, the River Samson, Belgium. *Hydrobiologia. 126:* 253-262.

Soszka, H. 1975. Chironomidae associated with pond-weeds *(Potamogeton lucens* and *Potamogeton perfoliatus)* in the Mikolajskie Lake. *Bull, de L'Acad. Sci 22:* 369-376.

Stimpson, K.S., Klemn, D.J. and Hiltunen, J.K. 1982. A guide to freshwater Tubificidae (Annelida : Clitellata : Oligochaeta) of North America. Environmental monitoring and support laboratory, Office of Research and Development, *U.S. Env. Protec.Agency,* Cincinnaty, Ohio. pp. 1-61.

Swink, U.D. and Novotny, J.F. 1985. Invertebrate colonization rate in the tailwater of a Kentucky flood control reservoir. *Jr. of Freshwat. Ecol. 3(1):* 27-34.

Thorp, J.H. and Chesser, R.K. 1983. Seasonal responses of lentic midge assemblage to envirornmental gradients. *Holartic Ecology. 6:* 123-132.

Tsui, P., Tripp, D. and Grant, W. 1978. A study to the biological colonization of the west Interceptor Ditch and lower Beaver Creek. Syncrude Canada Ltd., *Environ. Res. Monogr. 6:* 133-144.

Voigts, D.K. 1976. Aquatic invertebrate abundance in relation to changing marsh vegetation. *The American Midi. Nat. 95(2):* 313-322.

6

Effect of Endosulfan and Thimet on Histological Profile of Gills and Ovary of a Fresh Water Crab *Barytelphusa cunicularis*

T.J. Jadhav, G.K. Kulkarni[1]*, P.P. Joshi*[1] and *C.A. Jawale*[2]

SUMMARY

Changes induced in gills and ovary after exposure of crab *Barytelphusa cunicularis* to 0.14 ppm of endosulfan and 2.45 ppm of thimet 10 G were studied. The gills showed vacuolization in the gill stem, ruptured gill lamellae and damaged connective tissue cells in the stem. Branched erythrocytes in the stem and proximal lamellae were destructed and coagulation of haematocytes in the gill lamellae was noticed. The open faces of the lamellae which are joined by pillar cells were destructed. The lamellar tips which are devoid of pillar cells were also damaged. In the ovary the thin capsule of fibrous connective tissue enclosing the ovary was damaged. The outer thin epithelium and inner germinative epithelial layers were also damaged. The thin membrane covering the oocytes was also destructed and the follicle cells were atrophied. Vacuolation and fragmentation in the ooplasm were observed. Nutritive cells of oocyte were damaged. Endosulfan was found more toxic than thimet.

Department of Zoology, Shivaji College, Kannad (M.S.).

[1]Department of Zoology, Dr. Babasaheb Ambedkar Marathwada University, Aurangabad -43 1004 (M.S.).

[2]Department of Zoology, Shri Madhavrao Patil College, Murum-413 605 Dist. Osmanabad.

INTRODUCTION

Pesticides, because of their potential toxicity and indiscriminate usage, are known to produce morphological, anatomical and physiological changes in the vital organs such as reproductive, nervous, respiration osmoregulatory etc. of different non-target animals (Fingerman, 1982). Hazards due to the toxicants occur during manufacturing, transportation, storage, field use and contamination of water which ultimately find their way into aquatic ecosystem by many ways. There are several reports on toxic effects of pesticides on molluscs and crustaceans, in addition to fishes and frogs. Mane and Muley (1984) have determined the toxicity of endosulfan to two bivalve molluscs, *Lamellidens correanus* and *Lamellidens marginalis.* Mery et al. (1986) studied the size and sex dependent tolerance capacity to organophosphate pesticide in a fresh water prawn, *Macrobrachium kistnensis.* Thosar et al. (2000) observed changes in the structure of hepatopancreas of a freshwater snail *Viviparous bengalensis* exposed to sublethal concentration of metasystos. Pesticidal contaminants arising out of field, chemical industries and industrial effluents enter into aquatic ecosystem through rain runoff.

The respiration rate of organism is an indicative of the physiological state and change in respiration rates may be an indication of environmental stress (Capuzzo et al., 1976). Rao (1983) noticed vacuolization, necrosis and distention of cuticle in gill lamellage of crab, *Scylla serrata* after exposure to dimecron. Manikumar (1986) observed histological changes such as necrosis and vacuolization of gill tissue in prawn. *Macrobrachium lamerrii* after exposure to fenitrothion.

Reproduction is one of the essential activities of the organism. Effects of various pollutants and pharmaceutical drugs on invertebrate reproduction have been studied by some workers (Radhakrishnaiah and Renukadevi, 1990; Nowak and Babara, 1992; Lee and Oshima, 1998). However there are very few reports on effect of pollutants on histopathology of crustacean gonads (Bodkhe, 1983; Rao, 1983). Histopathological effects of various pollutants on crustancean reproduction has been studied by a few workers (Gyananath et al., 1987; Rao et al., 1987). Deecaraman and Subramoniam (1994) showed vacuolization and decrease in ooplasmic content in ovary and spermatheca of brackish water crab, *Uca triangularis benguli* after exposure to urea and naphthalin. Methyl parathion and phosalone induced several abonrmilities in the ovary of immature and maturing crab, *Scylla serrata* (Sharma et al., 1990). The time dependent toxicity of dimecron to the ovries of the atyid prawn, *Caridina rajadhari* has histologically been evaluated by Victor and Sarojini (1986).

In the present study the effect of endosulfan (organochlorine) and thimet (organophosphate) are studied on histological changes in the gills and the ovary of the crab *Barytelphusa cunicularis.*

MATERIALS AND METHODS

The freshwater crab, *Barytelphusa cunicularis* were collected from Daulatabad lake. The crabs were acclimatized to the laboratory condition for 2 days in plastic trough (18" diameter), 10 in each containing 2 litre of tap water before use for experiments. Healthy intermoult (stage C3) female crabs having approximately equal size (carapace width 35 to 40 mm) and weight (30±0.5g) were used for experiments. To study the histopathological lesions in the gills and ovary the crabs were exposed to 0.14 ppm of endosulfan 35 EC and 2.45 ppm of thimet 10 G for 24 h. (LC_{50} for 24 h.)

At the end of exposure period the experimental and control crabs were dissected and the gills and ovary were quickly excised and fixed in aqueous Bouins fluid. After fixation for 24 h the tissues were passed in 30% through 100% alcohol grades for dehydration and cleared in xylol. They were embedded in paraffin wax (m.p.58-60°C) and 7-8 µm serial sections were cut. The sections of gills and ovaries were stained in Delafield's haematoxylin and used eosin -Y as a counter stain (Bancroft and Stevens, 1982).

OBSERVATIONS AND RESULTS

Structure of the gills

Histologically the gills of *Barytelphusa cunicularis* show a middle gill stem which bears lamellae on both sides. There are three type of cells, (a) connective tissue cells in the stem, (b) branched arthrocytes in the proximal lamellae and (c) lamellar cells in the epithelium. Loosely packed arthrocytes, - blood spaces and occasional connective tissue fibres are seen in the stem. Epithelial cells of lamellae form the lining of gill stem. The individual lamella resembles a thin walled sac. The distal lamellae are expanded into a sac. At regular intervals, the open faces of the lamellae are joined by pillar cells. The lamellar tips are devoid of pillar cells. Each lamella consist of thin epidermal covering surrounding blood filled cavity known as lamellar sinus. The cavity is suppose to connect the afferent and efferent vessels in middle of the gill stem. A thin fluid band is present in between two gill lamellae. A thin membranous structure which covers the gill sac.

Effect of endosulfan and thimet on the gills

The gills of crab, *Barytelphusa cunicularis* exposed to 0.14 ppm of endosulfan for 24 h showed vacuolization in the gill stem, gill lamellae ruptured, connective tissue cells in the stem damaged, branched arthrocytes in the stem and proximal lamellae were destucted and coagulation of haemocytes in the gill lamellae was noticed. The open faces of the lamellae, which are joined by pillar cells are destructed. The lamellar tips which are devoid of pillar cells are also damaged. The thin connective fluidy band present in between the two gill lamellae was found ruptured. Comparison of results indicates that endosulfan has more deleterious effect than thimet.

Structure of the ovary

The histological structure of the ovary of control crab, *Barytelphusa cunicularis* showed that the entire ovary is enclosed by a thin capsule of fibrous connective tissue and associated cells. Microscopic observation of transverse sections of the ovary of freshwater crab, *Barytelphusa cunicularis* shows that it is covered with outer thin epithelium and inner germinative epithelial layer from which the oocytes proliferate. Thin membrane oocytes has large rounded nucleus with one or two nucleoli. The nutritive cells are present in close vacinity of oocytes and supply the nutritive material to the developing oocytes. The ooplasm is compactly arranged with thick yolk granules ovarian follicle are filled with different types of maturing oocytes. Fibrous connective tissue separates the lobes of mature ovaries. There are two types of cells in the ovarian lobes, the developing oocytes and the follicle cells. Oocytes are covered with a layer of follicle cells. Morphologically the, colour of the ovary also shows variation during the course of develpment: **(a) immature (stage I)** ovary is thin, white to pale yellow in colour, **(b) maturing (stage II)** ovary is dark yellow in colour, **(c) vitellogenic-I (stage III)** ovary is deep or dark yellow in colour. The ovarian lobes are large and extend on the abdominal regions, **(d) vitellogenic-II (stage IV)** ovary is fully matured and appears dark yellow in colour and (e) **spent resorptive stage (stage V**) ovary has aged oocytes almost of the same size that of the vitellogenic oocytes and colour is pale.

Effect of endosulfan and thimet on the ovary

The crabs exposed to 0.14 ppm of endosulfan and 2.45 ppm of thimet showed that the thin capsule of firbous connective tissue enclosing the ovary was destructed, the outer thin epithelium and inner germinative epithelial layer were damaged. The oocyte covering thin membrane was also damaged, and the follicle cells were destructed. Vacuolation and fragmentation in the ooplasm were observed pyconosis of nutritive cells and nucleus of oocytes. Endosulfan was found having more toxic effect than thimet

DISCUSSION

Changes in the histological structures are mainly directed to study the effect of pollutants on the structural components of cell. The potentially toxic pollutants induce morphological, behavioural and physiological changes in the vital organs (Fingerman, 1982). The present histopathological study on the freshwater crab, *Barytelphusa cunicularis* exposed to endosulfan and thimet shows severe changes in the gills structures which include distention of gill plates, vacuolisation and necrosis in gill stem and gill lamellae, the secondary lamellae was ruptured, the interlamellae distance was disturbed. Pillar cells were displaced with vacuolation in them. Marked degenerative changes were noticed in pillar cells, blood capillaries, epithelial cells and blood cells. Earlier Manikumar (1986) observed such histological changes in gills of the prawn, *Macrobrachium lamerrii* after exposure to fenitrothion. Various pesticides like endrin, aldrin, thiodon, etc. have been reported., to cause excessive mucus secretion, distintegration of cells, necrosis and damage of the respiratory epithelium of the gills of fishes and crabs (Das and Mukherjee, 2000; Jadav et al., 2004).

In the female crab *Barytelphusa cunicularis* maturing and mature oocyes were reduced in number, while atretic oocytes increased and damaged to the oocyte wall and the yolk vesicles are also seen. Generally, the organochlorine pesticides like endosulfan is more toxic and injurious than the organophoephate pesticide like thimet for crab *Barytelphusa cunicularis.* Effects of various pollutants on invertebrate reproduction have been studied by some workers (Radhakrishnaiah and Renukadevi, 1990; Nowak and Babara, 1992; Lee and Oshima, 1998). However, there are very few reports on effect of pollutants on histopathology of crustacean goands (Bodkhe, 1983; Rao 1983). Available report indicates that the pesticides affects the hormonal status of non-target species (Shanmugam et al:, 2000). When the toxic effect of pesticides is compared on the ovary of crab, *Barytelphusa cunicularis* the endosulfan is more toxic than thimet 10 G. Similar results were reported earlier by Sahai (1987) in a fish *Punctius ticto,* after exposing it to malathion, BHC, lindane and endosulfan and observed that these pesticides severely affected the oogenesis, induced necrosis of oocytes and vacuolization of cytoplasm and arrested vitellogenesis. Victor and Sarojini (1986) reported the toxicity of dimecron to the ovaries of the atyid prawn, *Caridina rajadhari* supporting the present observations.

REFERENCES

Bancroft J.D. and Steven A. (1982) Theory and practices of Histological Techniques. Churchil Livingstone. Publ. New York.

Bodkhe M.K. (1983) Effects of some pesticidal pollutants on the physiology of *Barytelphusa cunicularis.* Ph.D. Thesis. Marathwada University. Aurangabad. M. S. India.

Capuzzo J.M., Lawrence S. A. and Davidson T. A. (1976) Combined toxicity of free chlorine, chloramines and temperature to stage I larvae of the American lobster, *Homrus americanus.* Water Res. 10: 1093-1099.

Das B.K. and Mukherjee S. C. (2000) A histological study of carp *Labeo rohita* exposed to hexachlorocyclohexane. Vet. Archiv. 70 (4): 169-180.

Deecaraman M. and Subramoniam T. (1983) Synchronous development of the ovary and female accessory sex glands of crustacean *Squilla holoschista.* Proc. Indian Acad. Sci. (Anim. Sci.) 92 (2): 179-184.

Fingerman M. (1982) Key note address: Pollution: Our energy. All India Symposium on Physiological Responses of animals to Pollutants. Marathwada University, Aurangabad.

Gyananath G., Sarojini R. and Reddy T.S.N. (1987) Toxicity of potassium ferrocyanide and its sublethal effects on the ovary of the freshwater Prawn, *Macrobrachium lamerrii* Proc. Nat. Symp. Ecotoxic. 52-53.

Lee R. and Oshima Y. (1998) Effects of sublethal pesticides, metals and organometallics on development of blue crab *Callinectes sapidus* embryos. Mar. Environ. Res. 46 (1-5): 479-482.

Mane U.H. and Muley D.V. (1984) Acute toxicity of endosulfan 35EC to two freshwater bivalve mollusks from Godavari river at Paithan. Toxicol. Letters. 23: 147-155.

Manikumar D. (1986) Effect of pollutant on marine prawn. Ph.D. Thesis. Marathwada University, Aurangabad.

Maricharles P. (1988) Pesticide effect on the oogenesis of three freshwater prawns. Proc. 2nd Nat. Symp. Ecotoxicol. Abst. 10.

Mery Avelin Sr., Sarojini R. and Nagabhushanam R. (1986) Size and sex dependent capacity to organophosphate pesticides in freshwater prawn, *Macrobrachium lameri* Comp. Physiol. Ecol. III (14): 200- 202.

Nowak S. and Babara K. (1992) Histological changes in the gills induced by residues of endosulfan. Aquatic Toxico (AMST). 23 (1): 65-83.

Radhakrishnaiah K. and Renukadevi B. (1990) Size and sex related tolerance to pesticides in the freshwater field crab *Oziotelphusa senax senax.* Environ. Ecol. 1.8 (1A): 111-114.

Rao K. S. (1983) Effect of pesticidal pollutants on the reproduction and neurosecretion in marine edible crab, *Scylla serrata.* Ph.D. Thesis. Marathwada University, Aurangabad.

Rao K.S., Sarojini R. and Nagabhushanam R. (1987) Changes in the biochemical constituents of the marine edible crab, *Scylla serrata* after exposure to dimecron. Utter Pradesh J. Zool. 7 (1): 20-25.

Sahai S. (1987) Toxicological effects of some pesticides on the overies of *Punitius ticto.* Abst. 78. Proc. 8th AEB Soc. and Symp. Pollut. 124-132.

Shanmugam M., Venkateshwarlu-M. and Naveed A. (2000) Effect of pesticides on the freshwater crab *Barytelphusa cunicularis* (West Wood) J. Ecotoxicol. Environ. Monit. 10: 273-279.

Sharma H. A., Narabhushanam R. and Sarojini R. (1990) Effect of organophosphate insecticide on the edible crab *Scylla serrata.* Utter Pradesh J. Zool. 10 (2): 158-162.

Thosar M.R., Huilgoi N.V. and Lonkar A.M. (2000) Observations in the structure of hepatopancreas in the freshwater gastropod snail, *Viviparous bengalensis* (Lamark) exposed to sublethal concentration of insecticide metasystox. Internal. Conf. on Probing in Biological system. Abst. 118. p. 138.

Victor B. and Sarojini R. (1986) Toxicity of an organophosphorus insecticide to the ovaries of the Atyid prawn, *Caridina rajadhari.* J. Environ. Biol. 7(1): 47-50.

7

Effects of Mercuric Chloride and Copper Sulphate on Gonadal Indices and Gametogenesis of a Freshwater Leech, *Poecilobdella Viridis* (Blanchard)

S.D. Shelar[1], *G.K. Kulkarni*[2] *and G.V. Deshpande*[2]

SUMMARY

The reproductive process of leech shows interesting specialization when compared with those of other annelids. The leech *Poecilobdella viridis* is a protrandrous hermaphrodite, and bears high medicinal value as a source of anticoagulant substance from its salivary gland. It shares habitat with many other economically important animals like fishes, crustaceans, mollusks etc. in the freshwater ecosystem. It is non-target organism as far as effects of pesticides are concern. Pesticides, due to their potential toxicity and indiscriminate use are known to induce changes in the reproductive process of non-target organism like leeches. In the present study effects of mercuric chloride and copper sulphate on the gonadal indices and gametogenesis of the leech *Poecilobdella viridis* have been determined.

The leeches of equal size (7 ± 0.5 cm) and weight (50.0 ± 0.5 gm) were exposed to sub lethal concentration of mercuric chloride (0.2 ppm) and copper sulphate (0.6 ppm) for a period of 15 days during its pre-reproductive period (January). Slightly

[1]Department of Zoology, L.B. Shashtri College, Partur, Dist. Jalna.

[2]Department of Zoology, Dr. Babasaheb Ambedkar Marathwada University, Aurangabd-431 004 (M.S.).

clitellated leeches were selected. Gonadal indices were calculated using the formula of Farmanfarmaian et al., (1958) where as histological details of gonads were studied using double staining (Haematoxylene-nuclear stain and eosine-cytoplasmic stain) method of Bancroft and Stevens (1982).

Mercuric chloride produced 27% and 32% decreases in the testis and ovarian indices respectively, whereas 21% and 26% decreases in the testis and ovarian indices respectively were found in leeches treated with mercuric chloride and copper sulphate. Both of these heavy metals produced generalized inhibition of testis and ovarian maturation. Mercuric chloride was found to be more toxic than copper sulphate.

INTRODUCTION

The leech *Poecilobdella viridis* is a protandrous hermaphrodite (earlier development of sperms than ova) like temperate leech *Theromyzon rude* (Hagadorn, 1962). Extensive work has been done on the reproductive cycles, which are usually determined on the basis of gonadal indices, spawning potential, oocyte diameter frequency polygons and appearance of mature gametes etc.

Pesticides, because of their potential toxicity and indiscriminate use are known to produce morphological, anatomical and physiological changes in the vital organs such as reproductive, nervous, respiratory, osmoregulatory etc. of different non-target animals (Fingerman, 1982). Venna et al. (1992) suggested that malathion shows more toxic effects than copper on developing ovary of a fish *Etroplus maculatus*. Singh (1994) reported a decrease in the spermatogenesis in testes of a freshwater fish *Danio equipinatus* after exposure to inorganic salts. Aditya and Bandopadhyay (1995) observed that mercuric chloride treated planaria *Dugesia bengalensis* exhibits retarded growth and maturation. Van (1991) noticed that growth and sexual development was inhibited by cadmium, copper and pentachlorophenol in the earthworm *Eisenia ande* and suggested that from the ecological point of view growth and reproduction are more important for proper risk assessment of chemicals. Earlier Wills (1989) had noticed changes in shape and delay in deposition of cocoon in the leech *Erpobdella octaculata* after exposure to zinc. In the present study effects of mercuric chloride and copper sulphate on the reproductive activities of *Poecilobdella viridis* have been determined.

MATERIAL AND METHODS

The *Poecilobdella viridis* were collected from freshwater ponds around Aurangabad city and were maintained in glass troughs containing wet mud for 5 days to acclimatize them to the prevailing laboratory conditions under normal day night illumination (13 L: 11 D) at 27 ± 1°C. The leeches having approximately equal size (7 ± 05 cm) and weight (50.0 ± 0.5 gm) were selected for experimentation.

The lethal concentration of mercuric chloride and copper sulphate had been determined by static bioassay for a period of 96h (Shelar, 2002). To ascertain the effect of pesticides on reproductive organs, the leeches were collected during November to January (pre-reproductive period). The slightly clitelled leeches were selected for experiments. The leeches were divided in to different groups of 10 each and exposed them to sub lethal concentration of mercuric chloride (0.15 ppm) and copper sulphate (0.60 ppm) for a period of 15 days. The corresponding simultaneous control leeches were maintained in non-contaminated freshwater for the experimental period. After completion of the experiment the components of reproductive systems were dissected out and preceded as follows:

Reproductive organs like testis and ovary from 5 leeches of both experimental and control groups were dissected out and fixed in aqueous Bouin's fluid separately. After 24 h the fixed tissues were processed through dehydration in different grades of alcohol and embedded in paraffin-wax (m.p.58°C). Sections were cut at 7 µn thick and stained with Delafield's Haematoxylene with eosine-Y as the counter stain for histomorphological details.

From remaining 5 leeches the testes and ovaries were dissected out and gonad (testis and ovary) index was determined separately using the method of Farmanfarmaian et al. (1958) as follows:

$$\text{Gonad Index} = \frac{\text{Wet weight of a particular gonad}}{\text{Wet weight of leech}} \times 100$$

For every observation 100 randomly selected sections were examined and results are averaged.

OBSERVATIONS AND RESULTS

Gonadal index

The gonadal indices were calculated in relation to the total body weight. Testis index in the control leech was found to be 7.8±0.2 whereas ovarian index was found to be 1.1±0.4. Mercuric chloride produced 26.8% decrease in testis index and 31.58% decrease in ovarian index. Copper sulphate caused 21% decrease in testis index and 26.4% decrease in ovarian index (Table 7.1).

Gametogenesis

The percent occurrence of different stages of spermatogenesis and oogenesis is depicted in Table 2 and 3 respectively. The testis of simultaneous control leeches showed all the 4 stages of spermatogenesis such as immature (stage I), premature (stage II), maturing (stage III) and fully mature (stage IV) in a considerably high number. The pesticide treatment produced in general the inhibition of maturation of spermatogenic stages. In testis of mercuric chloride and copper sulphate treated leeches few stages of premature, maturing and fully mature oocytes were noticed but immature stages were still predominant (Table 2). All the four stages of oogenesis like undifferentiated (stage I), differentiated (stage II), previtellogenic (stage III) and vitellogenic (stage IV) were found in good number in the ovaries of simultaneous control leeches (Table 3). A few stage II, stage III but no stage IV oogonia were noticed in ovaries of mercuric chloride and copper sulphate treated leeches but stage I was still predominant.

DISCUSSION

The most authentically used method to study phase of reproductive cycles is the determination of indices of different reproductive organs. Belisle and Stickle (1978) reported that

Table 7.1: Percent decrease in the gonadal indices (against total body weight) of *Poecilobdella viridis* after mercuric chloride and copper sulphate treatments.

Treatment	*% Change in Testis index*	*% Change in Ovarian index*
Control	7.80 ±0.2*	1.10±0.4*
Mercuric chloride (0.1 5 ppm)	-26.00%	-31.5%
Copper sulphate (0.60 ppm)	-21.00%	-26.3%

* Original values.± Standard deviation from mean.

the index ratios provide a reliable knowledge of reproductive status of the animal. The reproductive cycle with oscillating trends in the histomorphological features of reproductive organs during different schedules of reproduction have been demonstrated in the earthworm *Perionyx excavatus* (Hanumante, 1975) and the slug *Laevicaulis alte* (Kulkarni, 1982). Usually the study of the reproductive organ index denotes the probable development and maturation of reproductive organs associated with different phase of reproductive cycle. The reproductive status of *Poecilobdella viridis* reflected through organ indexing method is associated with commencement and termination of reproductive cycle.

The result presented in Table 7.1 indicated that the gonadal indices decreased significantly after exposing leeches for 15 days to the sub lethal concentration of mercuric chloride and copper sulphate. Saxena and Agrawal (1991) while working on the toxicity of mercuric chloride on ovarian recrudescence in fish, *Channa punctatus* reported the low ovarian weight as well as retardation in the development of previtellogenic oocytes in to the vitellogenic yolk oocytes. Battacharya and Pandey (1989) reported significant fall in the gonadosomatic index in the fish, *Oreochromis mossambicus* exposed to an organochlorine, termax.

The obtained data show that the development and maturation stages of reproductive organs of *Poecilobdella viridis* decreased in general after exposure for 15 days to mercuric chloride and copper sulphate. The histological features of ovary (Table 7.2) and testis (Table 7.3) showed severe regression in maturation stages. Testis of simultaneous control leeches had mostly advanced stages of spermatogenesis like premature, maturing and fully mature with a few immature stages. Exposure of *Poecilobdella viridis* to mercuric chloride and copper sulphate for 15 days exhibit their profound effect not only on matured spermatogenic stages causing disturbance in the normal arrangement of spermatozoa around cytophore but also caused the disappearance of most of the other stages (Table 7.2). The histomorphological structure of the ovary of control leeches was appearing normal and contained more advanced stages (differentiated and a few previtellogenic) of oogenesis. The obtained results clearly show that there was a significant ($P < 0.05$) decrease in the number of differentiated stages after exposure to sub lethal concentrations of mercuric chloride and copper sulphate. They also induced drastic arrest of ovarian maturation and initiated degeneration of the ovarian cord (Table 7.3). These results clearly indicate that pesticides employed in this study have toxic effect on the gonads of *Poecilobdella viridis*. When the effect is compared, the mercuric chloride appeared to have more deleterious effect than copper sulphate. The toxicity of mercury to the ovaries of the prawn *Caridina rajadhari* has been histologically been evaluated by Sarojini and Victor (1985) and reported degeneration of oocyte, and extensive necrosis along with hypotrophy due to haematocyte accumulation. The effect of sodium selenite and its coexposure with cypermethrin on testis of adult grasshopper *Poecilocerus pictus* has been studied by Shyin and Usharani (1996) and reported that the increasing period of treatment resulted in pycnosis, spermatocyte loosening, sperm cells breakage of spermatozoa as well as spermatids. In the ovary, the zone of proliferation got affected and vacuolization in ooplasm and disappearance of nuclei in oocytes was evident. These results support the present findings in *Poecilobdella viridis*.

Table 7.2: Percent occurrence of different Spermatogenesis stages in the testes of *Poecilobdella viridis* after mercuric chloride and copper sulphate treatments.

Treatment	*% Spermatogenesis stages*			
	Immature (Stage I)	*Premature (Stage II)*	*Maturing (Stage III)*	*Fully mature (Stage IV)*
Control	65 ±6	23 ±4	11±2	1
Mercuric chloride (0.15 ppm)	63 ±5	21±3	11 ±2	5± 1
Copper sulphate (0.60 ppm)	58 ±5	26 ±2	11±2	7± 1

± Standard deviation from mean.

Table 7.3: Percent occurrence of different oogenesis stages in the ovaries of *Poecilobdella viridis* after mercuric chloride and copper sulphate treatments.

Treatment	*% Oogenesis stages*			
	Undifferentiated (Stage I)	*Differentiated (Stage II)*	*Previtellogenic (Stage III)*	*Vitellogenic (Stage IV)*
Control	69 ±6	22 ±4	6± 1	2±1
Mercuric chloride (0.15 ppm)	66 ±4	24 ±3	6± 1	2± 1
Copper sulphate (0.60 ppm)	62 ±3	29 ±3	4±2	2±1

± Standard deviation from mean.

Previous findings showed that gametogenesis is regulated by gonadotrophic hormone like principle from the brain of the temperate leech *Macrobdella decora* (Webb and Omar, 1981). The atresia of the ovarian cord and the regression of normal development of spermatozoa were supposed to be due to the lack of sufficient endogenous gonadotrophins, which are expected to be secreted by the central nervous system of the leech (Kulkarni and Nagabhushanam, 1980; Webb, 1980). Since profound changes in the gonads of *Poecilobdella viridis* after treatment with pesticides were observed in the present study it is quite possible that these pesticides might interfere with the gonadotrophin secretion and release, which is essential for normal gametogenesis (Mill, 1978). Loss of neurosecretory cell architecture of supra and suboesophageal ganglia of *Poecilobdella viridis* after exposing to mercuric chloride and copper sulphate has been reported earlier and suggested that these pesticides interfere with some of the vital biochemical processes of neuron which in turn make cell inactive (Shelar, 2002).

ACKNOWLEDGEMENT

We are thankful to the UGC, New Delhi for financial assistance through a project No. F3-60/2001 (SR-II). We appreciate technical support of Shri. Prashantkumar P. Joshi.

REFERENCES

Aditya A. and Bandopadhyay M. (1995) Mercuric chloride induced changes in *Dugesia bengalensis* (Karvakatsy), an aquatic planarian from Satinibetan, West Bengal. J. Environ. Biol. 16 (3): 233-236.

Agrawal S.K. (1991) Bioassay evaluation of acute toxicity levels of mercuric chloride to an airbreathing fish, *Channa punctatus*. J. Environ. Biol. 12 (2): 99-106.

Bancroft J.D. and Steven A. (1982) Theory and Practices of Histological Techniques. Churchil Livingstone. Publ. New York.

Battacharya L. and Pandey A.K. (1989) Inhibition of steroidogenesis and pattern of recovery in the testis of DDT exposed cichlid, *Orechromis mossambicus*. Bangaladesh. J. Zool. 17 (1): 1-14.

Belisle W.B. and Stickle W.B. (1978) Seasonal pattern in biochemical constituents and body components indices of muricid gastropod *Thais holmastoma*. Biol. Bull. 155:259-272.

Farmanfarmaian A., Giese A.C., Boolootian R.A. and Bennett J. (1958) Annual reproductive cycles in four species of West coast star fishes. J. Exp. Zool. 138:355-367.

Fingerman M. (1982) Key note address: Pollution: Our energy. All India Symposium on Physiological Responses of animals to Pollutants. Marathwada University, Aurangabad.

Hagadron I.R. (1962) Functional correlations of neurosecretion in the rhynchobdellid leech *Thermoyen rude*. Gen. Comp. Endocri. 21: 516-540.

Hanumante M.M. (1975) Some aspects of physiology of Indian earthworm. Ph.D. Thesis. Marathwada University. Aurangabad.

Kulkarni D.B. (1982) Studies on some endocrine aspects in the land slug, *Laevicaulis alte*. Ph.D. Thesis. Marathwada University, Aurangabad.

Kulkarni G.K. and Nagabhushanam R. (1980) Role of brain hormone in oogenesis of the Indian freshwater leech *Poecilobdella viridis* (Blanchard) during the annual reproductive cycle. Hydrobiologia. 69 (3): 225-228.

Mann K.H. (1962) Leeches (Hirudinea): Their structure, physiology, ecology and embryology: Interser monograph on pure and Appli. Sci. Zool. Div. II Pergamon Press, the MacMillan Company, New York. 1-201.

Mill P.J. (1978) Physiology of Annelids, Acadamic Press, New York.

Sarojini R. and Victor B. (1985) Toxicity of mercury on ovaries of the Caridian prawn, *Caridina rajadhari* (Bouvier) Curr. Sci. 54 (8): 398-400.

Shelar S.D. (2002) Physiological responses of a freshwater leech *Poecilobdella viridis* (Blanchard) to heavy metal pesticides. Ph.D. Thesis. Dr. Babasaheb Ambedkar Marathwada University, Aurangabad. M.S. India.

Shyin S and Usharani M.V. (1996) Histopathological effects of sodium selenate and co-exposure with cypermethrin on the testis of adult, *Poecilocerus pictus* (Orthoptera: Acrididae). J. Environ. Biol. 17(1): 11-15.

Singh A (1994) Oxygen consumption and ventilation rate in *Channa punctatus*.

(Bloch) exposed to sublethal levels of mercuric chloride. Env. Ecol. 12 (2) : 252-255.

Van Gastel C. A.M., Van Dis W.A., Dirven-Van Bremen E.M. Sparenburg P.M. and Bareselman R. (1991) Influence of cadmium, copper and pentachlorophenol on growth and sexual development of *Eisenia andrei*. Biol. Pert. Soil. 12 (2): 117-121.

Venna K.B., Radhakrishnan C.K. and Chacko T. (1992) Effect of copper and malathion on developing ovary of *Etroplus maculates* in the Cochin backwater. IInd Europen Conf. on Ecotoxico. 11-15, May, 1992. Abst. 44.

Webb R.A. (1980) Spermatogenesis in leeches. Evidence for a gonadotrophic peptide hormone produced by the supra oesophageal ganglion of *Erpobdella octoculata.* Gen. Comp. Endocrionol. 42: 401-412.

Webb R.A. and Omar E.F. (1981) Spermatogenesis in leeches (II); The effect of the supra oesophageal ganglion and ventral nerve cord ganglia on Spermatogenesis in the North American Medical Leech, *Macrobdella decora.* Gen. Comp. Endocrionol. 44: 54-63.

Wills M. (1989) Experimental studies on the effects of zinc on *Erpobdella octoculata* from the Afron Grafnant North Wales (UK). Arch. Hydrobiol. 116 (4): 449-476.

Bioaccumulation of Cypermethrin and Fenvalerate in Different Tissues of the Freshwater Fish, *Garra mullya* (Sykes)

P.P. Joshi and G.K. Kulkarni

SUMMARY

Bioaccumulation of persistent synthetic pyrethroid compounds results from the uptake and retention of residues in food and water. Cypermethrin is light stable and has a moderate persistence in soil. Fenvalerate is quite potent and stable to heat and moisture. Both acts as a contact and stomach poisons. Bioaccumulation of these pesticides was determined in gill, liver, intestine and muscle of a freshwater fish, *Garra mullya* (Sykes) after exposure to sub lethal concentrations 0.018 ppm of cypermethrin and 0.026 ppm of fenvalerate for a period of ten days. Maximum residue of 1.23 μg/g was found in gills followed by liver (0.96 μg/g), intestine (0.9 μg/g) and muscle (0.85 μg/g) in case of cypermethrin. Maximum accumulation of fenvalerate was found in gills (0.99 (μg/g) followed by liver (0.6 μg/g), muscle (0.45 μg/g) and intestine (0.4 (μg/g). The residue analysis revealed that cypermethrin is persistent and become accumulated to a greater extent in this fish than fenvalerate.

Key words: Cypermethrin, fenvalerate, bioaccumulation, fish *Garra mullya.*

Department of Zoology, Dr. Babasaheb Ambedkar Marathwada University, Aurangabad-431 004 (M.S.).

INTRODUCTION

The aquatic environment in recent times is witnessing an unprecedented impour of various kinds of biocides in alarming quantities and sources of such release are too many to be mentioned. With the onset of green revolution the problem of environmental contamination by excessive use of chemical pesticides was overlooked. Most of them are non-biodegradable; enhance their pollution potential and become a real danger to aquatic organisms because of its persistent nature in the environment (Murry and Beck, 1990; Bhavan et al, 1997). Pyrethroids are the new generation pesticides, which are good substitutes for organochlorines and organophosphates and are constantly gaining popularity (Susan, 1999). The knock down effect of pyrethroids on insects coupled with their extremely low toxicity to warm blooded animals made them suitable choice. They are more toxic to fish (Mushigeri and David, 2004; Prashanth et al., 2005). The pyrethroid pesticides are used in agriculture to control insect pest on various vegetables and cotton crops (Blossom, 2005). Cypermethrin is one of the recent pyrethroid being used for cotton pest control and wide variety of other crop pest in India. Fenvalerate is also recently developed type II synthetic pyrethroid has replaced other group of insecticide due to its improved potency and rapid biodegradability.

Fish are often considered to be much better and sensitive indicators of pesticide residues in the aquatic environment in which they are present. Bioaccumulation can be considered a sub lethal effect of pesticides but usually the organism evidences little impairment of normal activity or function. In the present study an attempt is made to know the persistence and bioaccumulation of cypermethrin and fenvalerate in different tissues of a freshwater fish *Garra mullya.*

MATERIAL METHODS

The freshwater fish *Garra mullya* were collected from Kham River at Tisgaon, Aurangabad district. They were acclimated to laboratory conditions (26 ± 1°C) for 15 days in a fifty liter capacity glass aquarium with sufficient quantity of tap water. During acclimation period the fishes were fed with commercially available standard palletized feed (Lipton India Ltd.). Aeration was provided. Commercial grade pyrethroid pesticide cypermethrin 25EC was obtained from Syngeta India Ltd. Mumbai and fenvalerate 20EC was obtained from Baroda Agrochemicals Ltd. Panelav Gujrat and used for the experiments. One ppm stock solution of pesticides was prepared in water. The healthy looking fishes of approximately equal weight (15±0.5g) and length (10±0.5cm) were selected for experimentation. The LC_{50} values for 24 h of cypermethrin and fenvalerate of the fish *Garra mullya* was already determined by Joshi and Kulkarni (2005). For bioaccumulation study the fishes were divided in to three groups of ten each. Static renewal bioassay method of Sprague (1973) was used to expose fishes to pesticides. Fishes of group one and two were exposed to sub lethal concentration ($1/10^{th}$. of 24 h LC_{30}) of cypermethrin (0.018 ppm) and fenvalerate (0.026 ppm) separately in glass aquaria of 50L capacity. Third was a simultaneous control group maintaining in tap water for a period of ten days. No aeration was provided during the experimental period. Test solutions were changed after every 24 h.

After the end of exposure period the fishes from all groups were taken out from respective aquarium one by one and dissected their different tissues viz. gills, liver, intestine and abdominal muscles. The tissues were collected in watch glasses, weighed and weighed and stored (<4°C) for subsequent analysis. The method of Salen et al. (1986) was used for extraction of residues from the tissue. The extract was cleaned up on silica gel columns covered with a layer of anhydrous sodium sulphate and packed with hexane (Goughan et al., 1978).

Gas liquid chromatographic analysis was carried out at Garware Analytical Center of Chemistry Department, Pune University on Hewlett-packard Model 5840, fitted with FID (Flame Ionization Dector) packed with 10% SE 30 on 80-100 mesh chromosorb WAW. The column temperature was 280°C. Injector and detector temperatures were 300°C respectively. Nitrogen was used as a carrier gas at a flow rate of 2.3 kg/cm^2. 5 µl of each sample was analyzed thrice with and without adding known amounts (100µg/g of sample) of cypermethrin and/or fenvalerate separately as an internal standard. Significance of the data was analysed using student 't' test (Mungikar, 2003).

RESULTS

Analysis of different tissues of fish *Garret mullya* following the exposure to sub-lethal concentrations of cypermethrin revealed that maximum residue of 1.23 μg/g was found in gills followed by liver (0.96 (μg/g), intestine (0.9 μg/g) and muscle (0.85 μg/g) (Fig. 8.1). The residue of fenvalerate was also found accumulated in all tissues of treated fish *Garra mullya.* The maximum accumulation of fenvalerate was found in gills (0.99μg/g) followed by liver (0.6 μg/g), muscle (0.45 μg/g) and intestine (0.4 μg/g) (Fig. 8.1).

DISCUSSION

Bioaccumulation is the ability of an organism to concentrate an element or a compound from food and surrounding water to a level higher than that of its environment (Menjer and Nelson, 1980). Thus, bioaccumulation studies are important in the estimation of potential to environmental harm. Gupta and Sharma (1994) reported on bioaccumulation of zinc in the fish *Cirrhinus mrigala* fingerlings during short term static bioassay. Athikesavan and Vincent (2000) studied toxic effects of nickel and zinc on bioaccumulation in *Hypophthalmichthys molitric.*

In the present study the accumulation of both cypermethrin and fenvalerate was found more in gills than other tissues. The degree of bioaccumulation of cypermethrin in various tissues of *Garra mullya* was in the order of: gills>liver>intestine>muscle, whereas in case of fenvalerate it was gills>liver>muscle>intestine. A similar observation was reported in freshwater prawn *Macrobrachium malcolmsonii* (Bhavan et al., 1997).

Susan et al. (1999) studied the bioaccumulation of fenvalerate in the whole body tissue of three major carps by GLC. They reported that the residues are highest in *Catla catla* and *Labeo rohita* exposed to sub lethal concentration for ten days. The isomer selectivity in aquatic toxicity and biodegradation of cypermethrin was also reported in *Ceriodaphnia* dubia by Liu et al. (2004).

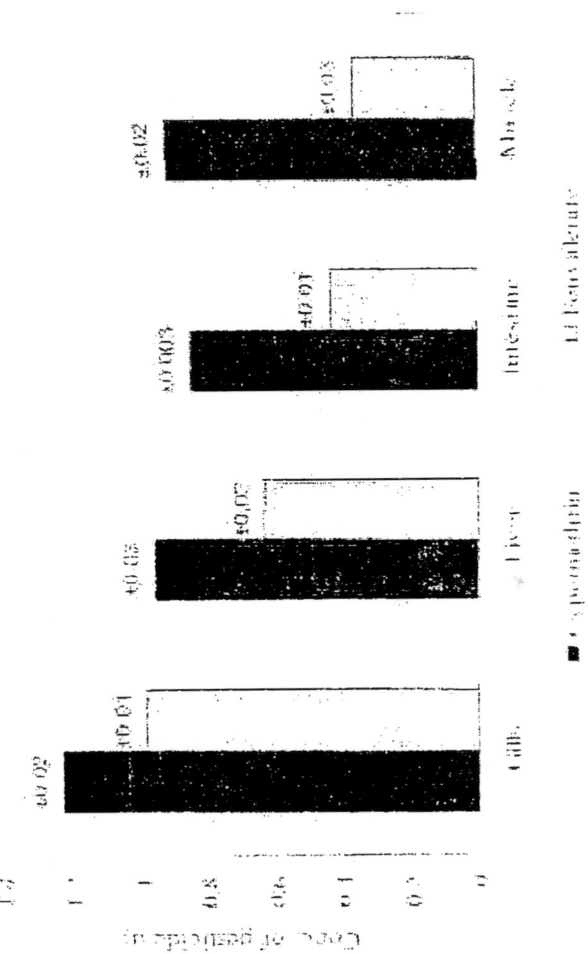

Fig. 8.1. Bioaccumulation of cypermethrin and fenvalerate in different tissues of freshwater fish *Garra mullya.*

Pesticides uptake and persistence depends on its solubility, interaction, chemical structure, fish conditions and variation in metabolic pattern (Johson, 1968). The variations observed in the tissues of *Garra mullya* could be due to sensitivity of the fish. A few reports are available on bioaccumulation of pesticides in aquatic animals such as in the freshwater fish, *Tilapia mossambica* (Thiangam and Sivakumar, 2004); *Rasbora daniconim* (Deshmukh, 2004); three species of *Labeo* (Saquib *et al.*, 2005) and freshwater field crab, *Spiralothelphusa hydrodroma* (Sreenivasan, 2005).

In conclusion, since pesticide residues are known to bioaccumulate in the tissues of fish and other animals and are also known to be transferred via food chain to the human bodies, the grave risk to the health of people who consume these fish seems to be considerable. The need to protect the people from undue exposure to the pesticide residues through the food chain can not be over looked. If the continuous discharge of these types of toxic pollutants is continued in to the river system it would adversely alter the physiology and quality of the fish, which ultimately affects its economic value.

REFERENCES

Athikesavan S. and Vincent S. (2000) Toxic effects of heavy metal pollutant nickel and zinc on bioaccumulation of freshwater fish *Hypophthamichthys molitrix.* Abst. 38. Natn. Sem. On Zool. For 21st Century, Goa Univ. From 7 to 9 December.

Bhavan S.P., Zayapragassarazan Z. and Geraldine P. (1997) Accumulation and elimination of endosulfan and carbaryl in the freshwater prawn *Macrobrachium malcolmsonii* (H Mile Edwards). Poll. Res. 16 (2): 113-117.

Blossom, Singh B. and Gaganjyot (2005) An appraisal of insecticide residues in lint and cotton seed in India. Pestology. XXIX (3): 53-57.

Deshmukh S.V. (2004) Persistence of copper in vital organs of freshwater fish *Rasbora daniconius* Comp. Toxicol. Physiol. 1 (II): 130-134.

Goughan L.C., Ackerman M.E., Uni T. and Casida J.E. (1978) Analysis of pesticide by G.L.C. J. Agric. Food Chem. 26: 613-615.

Gupta A.K. and Sharma S.K. (1994) Bioaccumulation of zinc in *Cirrhinus mrigala* (Hamilton) fingerlings during short term static bioassay. J. Environ. Biol. 15:231-237.

Johson B.T. (1968) Laboratory procedure for estimation of residue dynamics of xenobiotic contaminants in a freshwater food chain, Technical paper No. 103. Fish and Wild life Service, U.S. Department of the Interior, Washington, D.L.I.

Joshi P.P. and Kulkarni G.K. (2005) Toxicity and behavioural changes in freshwater fish *Garra mullya* exposed to cypermethrin and fenvalerate. Bioinfolet. 2 (3): 170-173.

Liu W., Gan J. J., Lee S. and Werner I. (2004) Isomer selectivity in aquatic toxicity and biodegradation of cypermethrin. J. Agric. Food Chem. 52:6233-6338.

Menjer R. E. and Nelson J. O. (1980) Water and soil pollutants In: D. J. Klasseiner and Amdu M. O. (eds). Toxicology, the basic Science of Poisons 7th edn. Mc. Millan. New York. 123-147.

Mungikar A.M. (2003) Biostatistical Analysis. Sarraswati Publ. Printing Press, Aurangabad (M.S) India.

Murry H.E. and Beck J.N. (1990) Concentration of selected chlorinated pesticides in shrimp collected from the Calcasieu river/lake Complex, Louisiana. Bull. Environ. Contam. Toxicol. 44: 798-804.

Mushigeri S. B. and David M. (2004) Toxicity of synthetic pyrethroid insecticide fenvalerate to the freshwater fish, *Cirrhinus mrigala*. J. Ecobiol. 16 (3): 169-174.

Prashanth M.S., David M. and Mathed S.G. (2005) Behavioural changes in freshwater fish, *Cirrhinus mrigala* (Hamilton) exposed to cypermethrin. J. Environ. Biol. 26 (1): 141-144.

Salen M.A., Ibrahim N. A., Soliman N. Z. and Seimy M. K. E. (1986) Persistence and distribution of cypermethrin, deltamethrin and fenvalerate in laying chickens. J. Agric. Food Chem. 34: 895-899.

Saquib T. A., Naqui S. N. H., Siddiqui P. A. and Azami M. A. (2005) Detection of pesticide residues in muscles, liver and fat of 3 species of *Labeo* found in Kalri and Haleji lakes. J. Environ. Biol. 26: (2 suppl): 433-438.

Sprague J.B. (1971) Measurement of pollution toxicity to fish II. Sublethal effects and safe concentrations. J. Water Research. 245-266.

Sreenivasan R.S. (2005) A study on the effects of cypermethrin a pyrethroid compound in *Spiralothelphusa hydrodroma*. Ph.D. Thesis. University of Madras. Chennai.

Susan A.T., Veeraiah K. and Tilak K.S. (1999) A study of the bioaccumulation of fenvalerate a synthetic pyrethroid in the whole body tissue of *Labeo rohita, Catla catla* and *Cirrhinus mrigala* (Hamilton) by GLC. Poll. Res. 18 (1): 57-59.

Thiangam R. and Sivakumar A. A. (2004) Bioaccumulation of chromium trioxide and its effect on blood glucose, glycogen content and LDH activity in the fish, *Tilapia mossambica*. Indian J. Environ. Ecoplan. 8 (2): 395-398.

Tilak K. S., Veeraiah K. and Sastry L.V. (2003) Bioaccumulation of fenvalerate technical grade in different organs of the frog *Haplobatrachus tigerinus* (Doudin) J. Environ. Biol. 24 (3): 261-264.

9

Water Quality Assessment of River Behta Using Benthic Microinvertebrates

M.P. Sharma, Shailendra Sharma, Vivek Goel Praveen Sharma and Arun Kumar

SUMMARY

Aquatic macroinvertebrates play significant role in responding to a variety of environmental conditions of rivers and streams and therefore may be used as bio-indicators for water quality assessment. In the past, biological communities like plankton, periphyton, microphytobenthos, macrozoo-benthos, aquatic macrophytes, fishes etc. have been used for the assessment of water quality of rivers and streams, but now the use of benthic macroinvertebrates as bio-indicators is gaining importance as these can be easily caught and seen with naked eyes and the method is less costlier and less time consuming compared to other methods given above.

Behta river of Paonta Sahib in Himachal Pradesh was chosen to assess the suitability of river water for drinking purposes. The present study involved sampling, pre-identification and identification of macroinvertebrates and computing the % occurrence of families of various taxonomic groups and conducting physico-chemical analysis of samples from selected location. Macroinvertebrates chosen were identified up to family level, and bioassessment at various locations has been done using NEPBIOS score system. It was found that out of total 30 genus belonging to 10 families of taxonomical group

Alternate Hydro Energy centre Indian Institute of Technology, Roorkee 247667, Uttaranchal.

like *Ephemeroptera, Trichoptera, Plecoptera, Coleoptera, Heteroptera, Odonata, Diptera Mollusca, Oligochaetes* etc. have been found in different composition inhabiting the river. The results further show that all the locations assessed for quality using macroinvertebrates and physico-chemical analysis are in the range of water quality class III (Moderately Polluted) and the water can not be used for drinking purposes. The measures to reduce point and non-point sources of pollution have been suggested to get the quality suitable for drinking purposes.

INTRODUCTION

Aquatic macroinvertebrates play significant role in responding to a variety of environmental conditions of rivers and streams and therefore may be used as bio-indicators for water quality assessment. Benthic macroinvertebrates are the animals that lake a backbones and generally visible to the naked eyes. They live in the lower areas of the streams under rocks.They include larval forms of many common insects such as Dragon flies, Damsel flies and Crane flies. Common features of these are as follows:

- Live in water for all or most of their life. Often live for more than on year.
- Stay in the area suitable for their survival.
- Differ in their tolerance to amount and types of pollution.
- Are easy to identify in the laboratory.
- Have limited mobility.

Macro-invertebrate community responses to environmental changes are useful in assessing the impact of municipal, industrial and agricultural waste and impacts from other land uses on surface water. The macroinvertebrates are highly popular as pollution indicators (Hallaswill, 1986).

Benthic Organisms are of great significance because they form the food of fishes and their productivity play an important link in the food chain, Benthic Organisms are distrivores and form an important link in the food chain, an account of their ability to convert low quality and low energy detritus into better quality food for higher organisms in the food web with the unfolding of the importance of benthos in food chain, benthic productivity has been correlated with fish resources. The qualitative and quantitative changes in the benthic population have also been used as pollution indices, (Oommachan 1981; Bhutiani 2004; Kumar 2003; Tyagi et al., 2006). Benthic macroinvertebrates are aquatic macrofauna inhibiting the bottom substrate for at least a part of their life cycle. The reason of selecting macroinvertebrates as bio-indicators are, that they are visible to unaided eyes and retained on the sieve with a mesh sized of 0.6 mm diameter. They have sedentary and long life span and sensitive community response to organic loading, thermal impacts, substrate alteration and toxic pollution. Inhabiting the different substratum of river, stream, lake and other water bodies, developed taxonomy and integrated of pollution etc. justifies the reason of selecting them as bio-indicators.

Therefore the primary objective of study was to evaluate the water quality of river Behta for drinking purposes using macroinvertebrates. The other objectives were to describe the importance of using macroinvertebrates as pollution indicator and the bioassessment result validation by physico-chemical analysis using National Satiation Foundation water quality index. Also discussed the occurrence of benthic macro-invertebrates community along with the distribution of taxa-group of the river has been discussed.

ABOUT BEHTA RIVER

The Behta river is an important tributary of river Yamuna. It's originates in the boulders below the Nahan ridge in the South-Western corner of Himachal Pradesh as the Jalmusa-Ka-Khala (Fig. 9.1). Behta river of Paonta Sahib in Himachal Pradesh was chosen to assess the suitability of river water for drinking purposes.

Physico-Chemical and Biological parameters for two sites on the river Behta at Paonta Sahib were analyzed,and the results revealed that the water quality at the site u/s to slaughter house was good which belong to water quality class II, and the water quality at the site D/S to slaughter house was Moderate which belong to water quality class III. The conclusion of the results is that the water at the D/S of the Slaughter house can not be used for the drinking purposes. This river is mainly fed by the rain water that is cycled as underground water before finally coming up on the surface as a spring. The river flows below the surface for a part of its length in its upper reaches, thereafter the water flows on the surface.

There were two sampling sites selected by us in Paonta Sahib (H.P.):

i) River Behta at Paonta Sahib U/S to Slaughter house (Station-01) at Longitude 77.55, Latitude - 30.47 and Altitude (m) - 380.0. (Fig. 9.1)

ii) River Behta at Paonta sahib 500 m D/S to Slaughter house (Station - 02) at Longitude 77.57, Latitude - 30.44 and Altitude (m) - 369.0. (Fig. 9.1)

The sampling sites are situated within a landscape characterized by cropland, clear cutting, urban sites and industrial activities. 500 m above the sampling site there is a chicken farm. The riverbed is build by meso- and micro-lithal 60% and 40% respectively. Filamentous algae and algae tufts are occurring frequently. The average stream width is up to 35 m, mean depth is 40 cm and mean current velocity is 25 cm/s. The water carries foam and is turbid. Mud and stones show reduction phenomena both in lentic and lotic areas.

METHODOLOGY

A sample consists of collection of 20 sub-samples each of (0.25 x 0.25) m^2 taken from all microhabitat types. This procedure (Fig. 9.2) results in sampling of approximately 1.25 m^2 stream bottom area. Net of mesh size 500 μm is used for collecting the macro invertebrates. Every large boulder or cobble in the area is picked up if it could be lifted and organisms vigorously washed by hand into the net. Finally, the substrate with smaller boulders should be disturbed by kicking systematically across the area 3-4 times such that the invertebrates wash downstream into the net. The organisms are then carefully

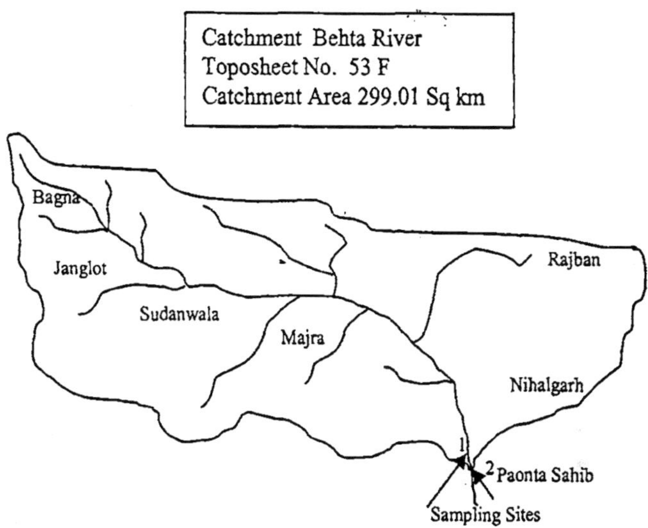

Fig. 9.1 : Catchment area of river Behta at Paonta sahib

picked from the net surface and preserved immediately in 80% ethanol or 4% formaldehyde. These samples are returned to the laboratory for processing. Specimen collected are sorted and identified to operational taxonomic unit (at least to family level with the help of regional keys) in the laboratory under a dissecting microscope for identifying the fauna, standard literature was consulted (Needhan & Needham (1988), Pennak (1989), Tonapi (1980), Williams & Feltmate 1992 Merrit & Cummins 1984). The results of the analysis are reported in Table 9.2.

Samples for Microbiological examination were collected in Non-reactive Borosilicate glass bottles that have been cleansed and rinsed carefully, given a final rinse with the distilled water and sterilized in Autoclave.

Water samples were collected in plastic container for different Physico-Chemical Parameters. The chemical characteristics were determined by the standard methods suggested of APHA (1998), (The results of the analysis are reported in Table 9.1.)

The schematic flow chart of the steps involved in the Methodology is given as below:

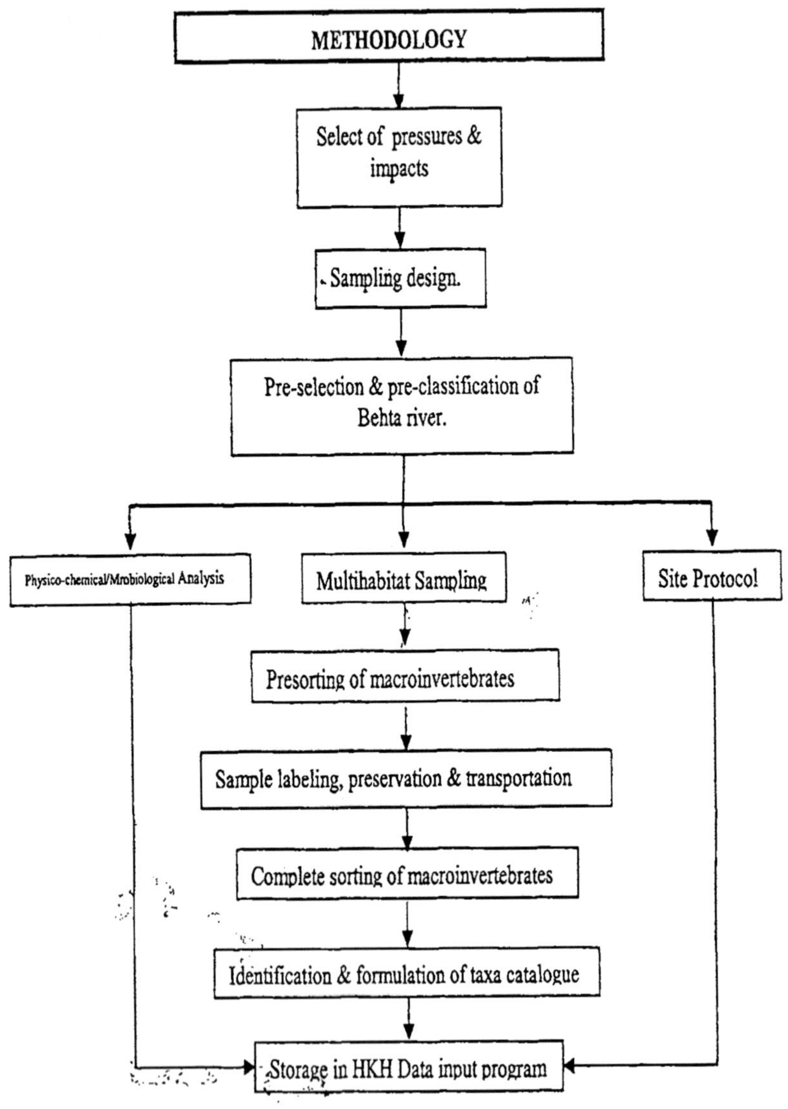

Fig. 19.2 Flow chart of Methodology

RESULTS AND DISCUSSION

(A) Physico-Chemical and Biological Parameters

Following parameters were analyzed -

Table 9.1 : Physico-chemical and Biological analysis of water samples.

Site Name	*Station-01*	*Station-02*
Pre-classification Class	2	3
Estimated discharge [l/s]	510.6	511
Temp. (Water) °C	26.5	26.5
Temp. (Air) °C	37.0	37
pH	7.84	8.2
Conductivity [μS/cm]	280.0	366
Turbidity NTU	0.84	1.42
Oxygen content [mg/l]	9.54	9.09
% Saturation of Oxygen	109.4	108.3
Alkalinity [CO_3 2-] [mmol/l]	126.0	141
Total Hardness [mmol/l]	197.0	201
Chloride [mg/l]	12.2	13.4
Ammonium [mg/l]	0	0
Nitrite [mg/l]	0.00,1	0.004
Nitrate [mg/l]	0.28	0.3
Ortho -phosphate [Mg/l]	98.0	110
Total phosphate [μg/l]	701.0	940
BODS [mg/l]	2.1	3
E-coli [n/100 ml]	500	1600
TDS [ppt]	0.16	0.22
Estimated class using NSF W.Q.Index	II	III
NSF Index Value	75	69
Water Quality Index Legend	71-90	50-70

On the basis of these chemical parameters the Water quality can be determined using NSF water quality index.

NSF water quality index:

NSF International, founded in 1944 as the National Sanitation Foundation, is known for the development of standards, product testing and certification services in the areas of public health, safety and protection of the environment.

The index is basically a mathematical means of calculating a single value from multiple test results. The index result represents the level of water quality in a given water basin, such as a lake, river, or stream.

The Water Quality Index uses a scale from 0 to 100 to rate the quality of the water, with 100 being the highest possible score. Once the overall WQI score is known, it can be compared against the following scale to determine how healthy the water is on a given day.

WQI quality scale

91 -100	:	Excellent water quality
71-90	:	Good water quality
51-70	:	Medium or average water quality
26-50	:	Fair water quality
0-25	:	Poor water quality

Water supplies with ratings falling in the good or excellent range would able to support a high diversity of aquatic life. In addition, the water would also be suitable for all forms of recreation, including those involving direct contact with the water. Water supplies achieving only an average rating generally have less diversity of aquatic organisms and frequently have increased algae growth.

Water supplies falling into the fair range are only able to support a low diversity of aquatic life and are probably experiencing problems with pollution. Water supplies that fall into the poor category may only be able to support a limited number of aquatic life forms, and it is expected that these waters have abundant quality problems. A water supply with a poor quality rating would not normally be considered acceptable for activities involving direct contact with the water, such as swimming.

Station 1 : Behta at Paonta Sahib U/S to Slaughter house

pH	*Total Solids (ppt)*	*BOD Mg/l.*	*DO%*	*Nitrates Mg/l.*	*Phosphates Mg/l.*	*Turbidity NTU*	*E.Coli N/100ml.*
7.84	0.16	2.1	109.4	0.28	0.70	0.84	500

The range using NSF water quality index for the site is 75, which is indicative of good water quality (Class II).

Station 2 : Behta at Paonta Sahib 500 m D/S to Slaughter house

pH	*Total Solids (ppt)*	*BOD Mg/l.*	*DO%*	*Nitrates Mg/l.*	*Phosphates Mg/l.*	*Turbidity NTU*	*E.Coli N/100ml.*
8.2	0.22	3.0	108.3	0.3	0.94	1.42	1600

The range using NSF water quality index for the site is 69, which is indicative of moderate water quality (Class III).

(B) Biological Parameters

The common and dominant families of macroinvertebrates of each group encountered are as follows:

The calculation of water quality on the basis of macroinvertebrates families was done on the basis of NAPBIOS using pre-classification Sheet.

NEPBIOS Biotic Index

A suitable biological method based on indices or score system is possible only when local reference communities are properly scored. Taking this fact into consideration the Nepalese taxa were scored, the average score per taxon (ASPT) calculated and a different biotic score method for Nepal developed with the name Nepalese Biotic Score abbreviated as NEPBIOS (Table 9.3). The calculation of the water quality on the basis of the presence of the macroinvertebrates families is done on the basis of NAPBIOS using the Pre-classification sheet.

Table: 9.2: Occurrence of Benthic Macroinvertebrates community

Taxonomic Group		*Family*	*Behta at Paonta Sahib 10 m U/S to Slaughter House. (Station-01)*		*Behta at Paonta Sahib 500 m D/S to Slaughter (Station-02) house*	
			No. of individual	*% of Abundance*	*No. of Ind.*	*%*
Mollusca	1-	*Thiaridae*	431	25.11%	1951	67.67%
	2-	*Planorbidae*	—	—	4	0.13%
	3-	*Lymnaeidae*	2	0.11%	4	0.13%
	4-	*Pisidium*	—	—	1	0.03%
	5-	*Viviparidae*	4	0.23%	—	—
Epheneroptera	1-	*Neoephemeridae*	13	0.75%	1	0.03%
	2-	*Baetidae*	617	35.95%	324	11.23%
	3-	*Ephemerilidae*	31	1.80%	2	0.06%
	4-	*Heptageniidae*	52	3.03%	10	0.34%
	5-	*Ephemeridae*	23	1.34%	—	—
	6-	*Leptophlebiidae*	12	0.69%	3	0.10%
	7-	*Caenidae*	17	0.91%	1	0.03%
Odaonata	1-	*Gomphidae*	11	0.64%	76	2.63%
	2-	*Libellutidae*	3	0.17%	1	2.67%

Taxonomic Group		*Family*	*Behta at Paonta Sahib 10 m U/S to Slaughter House. (Station-01)*		*Behta at Paonta Sahib 500 m D/S to Slaughter (Station-02) house*	
			No. of individual	*% of Abundance*	*No. of Ind.*	*%*
Colleptera	1-	*Elmidae*	21	1.22%	1	0.03%
	2-	*Hydroptillidae*	8	0.46%	3	0.10%
	3-	*Dryopidae*	29	1.68%	—	—
	4-	*Psepheniidae*	59	3.43%	7	0.24%
Trichoptera	1-	*Hydropsychidae*	53	3.08%	110	3.81%
	2-	*Glossosomatidae*	63	3.6%	6	0.20%
	3-	*Lepidostomatiodae*	12	0.69%	—	—
	4-	*Polycentropodiae*	19	1.10%	9	0.31%
	5-	*Hydroptillidae*	32	1.86%	4	0.13%
	6-	*Wenoidae*	17	0.99%	—	—
	7-	*Leptoceredae*	3	0.17%	1	0.03%
	8-	*Rhyacophillidae*	13	0.75%	—	—
	9-	*Philopotamidae*	17	0.99%	2	0.06%
	10-	*Goeridae*	34	1.98%	—	—

Taxonomic Group		*Family*	*Behta at Paonta Sahib 10 m U/S to Slaughter House. (Station-01)*		*Behta at Paonta Sahib 500 m D/S to Slaughter (Station-02) house*	
			No. of individual	*% of Abundance*	*No. of Ind.*	*%*
Diptera	1-	*Tabanidae*	3	0.17%	13	0.45%
	2-	*Chironomidae*	11	0.64%	55	1.90%
	3-	*Tipulidae*	—	—	9	0.31%
	4-	*Ephydridae*	3	0.17%	13	0.45%
	5-	*Simulidae*	19	1.10%	2	0.06%
	6-	*Ceratopogonidae*	—	—	1	0.03%
Hemiptera	1-	*Corixidae*	22	1.28%	3	0.10%
Placoptera	1-	*Perlidae*	36	2.01%		.
Crustacea	1-	*Palaemonidae*	23	1.24%	176	6.10%
Annelida	1-	*Oligochaetes*	3	0.17%	89	3.08%

Table 9.3: Nepalese Biotic Scores (NEPBIOS) assigned to the Macroinvertebrates.

S. No.	*Macroinvertebrates*	*Score*
1.	Capniidae, Ephemerellidae *(Drunella* sp.), Epiophlebiidae, Helicopsychidae, Helodidae (Scirtidae), Heptageniidae *(Epeorus rhithralis),* Heptageniidae *(Rhithrogena nepalensis),* Leuctridae, Peltoperlidae, Perlidae *(Acroneuria* spp.), Perlidae *(Calicneuria* spp.), Siphlonuridae, Taeniopterygidae, Uenoidae.	10
2.	Athericidae, Chloroperlidae, Goeridae, Leptophlebiidae *(Habrophlebiodes* sp.), Limnocentropodidae, Neoephemeridae, Perlodidae, Polycentropodidae.	9
3.	*Baetidae (Centroptilumsp.),* Brachycentridae, Chironomidae (Diamesinae), Elmidae, Euphaeidae, Glossosomatidae, Heptageniidae *(Epeorus bispinosus),* Heptageniidae *(Iron* psi), Heptageniidae *(Rhithrogena spp.),* Hydrobiosidae, Lepidostomatidae, Limnephilidae, Nemouridae, Pdrlidae, Philopotamidae, Psephenidae, Rhyacophilidae, Stenopsychidae.	8
4.	Aphelocheiridae, *Baetidae (Cloedodes sp.),* Baetidae *(Baetiella spp.),* Baetidae *(Baetis spp.),* Baetidae *(Baetiella ausobskyi), Baetidae (Baetis sp.l),* Corydalidae, Ephemerellidae, Ephemerellidae *(Cincticostella sp.),* Ephemeridae, Gammaridae, Gyrinidae, Heptageniidae, Heptageniidae *(Cinygmina* sp.), Heptageniidae *(Notacanthurus cristatus),* Hydraenidae, Leptophlebiidae, Limoniidae, Pleuroceridae, Psychomyiidae, Salifidae *(Barbronia sp.),* Simuliidae, Tipulidae.	7
5.	Aeshnidae, Baetidae *(Baetis sp.5), Baetidae (Baetis sp.4),* Caenidae, Ceratopogonidae, Ecnomidae, Ephemerellidae *(Torleya nepalica),* Heptageniidae *(Electrogena* sp.), Hydrometridae, Hydropsychidae,	6

	Hydroptilidae, Potamidae, Scirtidae, Viviparidae.	
6.	Baetidae *(Baetis sp.2)*, Baetidae *(Baetis sp.3)*, Bithyniidae, Chlorocyphidae, Coenagrionidae, Corduliidae, Dryopidae, Hydrophilidae, Leptophlebiidae *(Euthraulus spp.)*, Lymnaeidae, Odontoceridae, Protoneuridae, Sphaeriidae, Unionidae.	5
7.	Calopterygidae, Chironomidae *(Microtendipes* sp.), Chironomidae *(Polypedilum* sp.), Corbiculidae, Dytiscidae, Gerridae, Glossiphoniidae, Micronectidae, Naucoridae, Nepidae, Palaemonidae, Planorbidae, Ranatridae, Salifidae *(Barbronia weberi)*, Thiaridae.	4
08.	Corixidae, Libellulidae, Lumbricidae, Noteridae, Notonectidae, Salifidae	3
09.	Culicidae, Physidae, Tubificidae	2
10.	Chironomidae *[Chironomus* group *riparius (=thummi)* and group *plumosus]*	1

Three procedures were followed in scoring the taxa.

(1) *Numerical Procedure:* This procedure follows the following formula:

Guide Score = SI/STot × 10 + SI-II/STot × 8.57 + SII/STot × 7.14 + SH-III/STot × 5.71 + SIII/STot × 4.28 + SIII-IV/STot × 2.85 + SIV/STot × 1.43

Where,

SI, SI-II, SII, SII-III, SIII, SIII-IV, SIV are the total number of sites representing the pollutional classes I, I-II, II, II-III, III, III-IV, IV.

Stot = SI+SI-II+SII+SII-III+SIII+SIII-IV+SIV

1.43 is the score interval with 10 as maximum.

(2) *Professional judgements:*

Step-I: Based on the reference made to the scores that has previously been assigned by different authors in their respective country of origin, and the range of pollutional class represented by each taxon in the rivers of Nepal.

Step-II: Based on the distribution pattern of each taxon (family level) in response to pollutional level. The comparison of family (taxon) distribution with the distribution of observed water quality classes was carried out to observe if any families with the same ecological distribution are differently scored. If so whether or not the reasons are matching.

Once NEPBIOS/ASPT is calculated reference should be made to the transformation table below for interpretation on the water quality of the particular site investigated (Table 9.4).

Table 9.4: Water quality scores based on NEPBIOS.

NEPBIOS/ASPT	*Water quality*
8.00-10.00	I
7.00-7.99	I-II
5.50-6.99	II
4.00-5.49	II-III
2.50-3.99	III
1.01-2.49	III-IV
1	IV

On the Basis of the NEPBIOS score system, the species present in the samples of the site at Behta River 10 m U/S to the slaughter house shows that water quality of the river belongs to class-II.

On the Basis of the NEPBIOS score system, the species present in the samples of the site at Behta River 500 m D/S to the slaughter house shows that water quality of the river belongs to class-III.

During the investigation at station-01, it was found that water quality was good with the pH of 7.84 and turbidity of 0.84NTU. The Biological Oxygen Demand is 2.1. Total 34 families of Macroinvertebrates belonging to Groups: **Ephemeroptera, Coleoptera, Trichoptera, Diptera, Plecoptera, Hemiptera, Crustacea, Annelida, Mollusca and Odonata** were encountered. The insect population represented 72.98% of total fauna and belonging to orders **Trichoptera, Ephemeroptera, Coleoptera, Diptera, Odonata and Hemiptera.** The order **Ephemeroptera, Trichoptera, Coleoptera and Placoptera** are dominating in numbers. The results further show that all the locations assessed for quality using macroinvertebrates and physico-chemical analysis are in the range of water quality class II (Good) and the water can be used for drinking purposes.

During the investigation at station-02, it was found that river water was a little alkaline with pH of 8.2 and moderately polluted with turbidity. The study of fresh water macroinvertebrates shows that 30 Families belonging to groups **Mollusca, Odonata, Ephemeroptera, Coleoptera, Trichoptera, Diptera, Plecoptera, Hemiptera, Crustacea and Annelida** occurred in the river.

The insect population represented 22.53% of total fauna of Behta river and was belonging to the order **Odonata, Ephemeroptera, Coleoptera, Trichoptera, Diptera, Placoptera, Hemiptera.** Insect have the capability to adapt to varied aquatic habitats due to their extra ordinary structural organization (Tonapi (19801), Needham & Needham (1988), Tyagi et al (2006). The benthic population of aquatic insects was dominated by **Trichoptera** comprising 8 families and diptera comprising 6 families. Most of these families to be tolerant to varied aquatic environment (Williams) & feltmate 1992, Manit & Cummines (1984). Bath & Kaust (1997).

The **Mollusca** fauna of Behta river was represented by 4 families out of which **Thiaridae** family dominated the population. Covers 67.82% of the total population of aquatic fauna. This group has significant positive correlation with the Total hardness (201.0 mmol/l), Alkalinity (141.0 mmol/l), Phosphate (0.94 mg/l) and Chloride (13.4 mg/l). The rest of the aquatic invertebrate fauna of behta river of one family of Crustacea, and a Annilida.

CONCLUSION

Benthic macroinvertebrates community as a whole in the river has been found to have significant positive correlation with the Total hardness, Total Alkalinity, Chloride, Phosphate and Transparency.

The results show that all the locations assessed for quality using macroinvertebrates and physico-chemical analysis are in the range of water quality class III (Moderately Polluted) at the station II and the water can not be used for drinking purposes. The measures to reduce point and non-point sources of pollution have been suggested to get the quality suitable for drinking purposes.

ACKNOWLEDGEMENT

The authors are thankful to ASSESS-HKH (Project No. 003659) for financial assistance to carry out the work which is a part of program of developing an assessment tool.

REFERENCES

APHA (1998), *Standard method for the examination of water and waste water*, American Public Health Association, 20th Ed. New York.

Arvind Kumar, 1994: *Role of species diversity of aquatic insects in the assessment of population in wetlands of Santhal Parganas (Bihar). J. Environment and Pollution* (1, 3 & 4). 117-120.

Bath, K.S. & Kaur, 1997. *Aquatic insects as bio-indicators at Harike reservoir in Punjab-India. Indian journal of Environmental Sciences* 2: 133-138.

Bhutiani, R. 2004. *Limnological Status of River Suswa with reference to its Mathematical Modelling*. Ph.D. Thesis, Gurukul Kangdi. Vishwavidyalaya, Haridwar.

Dhillon S.S. Bath, K.S. and Mander, G. 1995. *Invertebrate fauna of fresh waterbodies existing in and around Patiala. Journal of Environment and Pollution* 2:163-167.

Hach Water testing kit methods (USEPA accepted for reporting for waste water analysis-2004).

Hinger L.W.C. (1992) *A checklist of the trichoptera recorded from India & a larval key to the families; Oriental Insects.*

Kumar K (2003) *Bioassessment of water quality of river Yamuna using Benthic macroinvertebrates:* M.Sc. Thesis Delhi University.

Malhotra Y.R., Gupta, K. and Khajuria, A. 1990. *Seasonal variation in the population of macro-zoobenthos in relation to some physico-chemical parameters of lake Mansar.J. Freshwater Biol.* 2: 123-128.

Merrit, R.W. and K.W. Cummins, eds. 1984. *An Introduction to the aquatic insects of North America.* 2nd ed. Kendall/Hunt Publishing Company, Dubuque.

Needham, James C. and Paul R. Needham. 1988. *A Guide to the study of fresh-water Biology. Reiter's Scientific and professional books,* Washington, D.C.

Pennak, Robert W. 1989. *Fresh-water invertebrates of the United States: Protoza to Mollusca.* 3rd. ed. John Wiley and Sons, New York.

Roy, S.P. and Sharma, U.P. 1983. *Dipteran larval population in a fresh-water wetland at Bhagalpur (Bihar) in relation to physico-chemical parameters. Ind. J. Ecol.* 10:129-136.

Srivastava, H.N. 1962. *Aquatic fauna as indicators of Pollution. Env. Hlth.* 106-113.

Subba Rao, N.V. (1993): Freshwater Molluscs of India in *Recent Advances in freshwater Biology* (ed. KS. Rao) 187-220, 4+61.

Tonapi, G.T. 1980. *Fresh water animals of India-An ecological approach.* Oxford and IBM Publishing Co. New Delhi.

Tyagi,P.et al(2006) *Occurrence of benthic macroinvertebratres families encountered in river Hindan in Uttar Pradesh (India).J.Zool.India,* Vol. 9, No. 1, pp. 209-216.

Williams, D.D. & Feltmate, B.W. 1992. *Aquatic insects. C.A.B. International,* Wallingford, Oxon, UK.

Seasonal Variations in Some Abiotic Factors of Moghat Reservoir Khandwa (M.P.)

KESHRE VIVEK AND L.K. MUDGAL***

SUMMARY

Khandwa is one of the district of the state M.P. The Moghat reservoir is situated three kilometers away in the North East area of Khandwa town on 21°49'36" N latitude and 76°20'56" E longitude. It is a man made reservoir built in 1897. The present study aims at making an assessment of the water quality of Moghat reservoir of Khandwa M.P.

Introduction : The most important gift for mankind is the water which plays a significant role in different vital and structural activities. Water is inevitable for all living organisms as it has a great social and economical value ultimately affecting mans health. It is essentially required for industrial development, Fisheries, Irrigation hydro electrical generations human life survival and domesticated animals.

Water covers more than 75% of the earth surface either as salt water or fresh water. It is an excellent universal solvent as well as buoyant medium. Utility of water for multiple purposes depends upon its quality. The quality of water is primarily based upon its physical chemical and biological characteristics. The present study aims at making an assessment of the water quality of the Moghat reservoir of Khandwa M.P.

*Govt Girls College Khandwa (M.P.).

**Govt Girls College Motitabela Indore(M.P.).

Khandwa is one of the district of Madhya Pradesh which is famous for its historical and holly place Omkareshwar where one of the Jyotirlinga of Lord Shiva is situated on the bank of river Narmada. On the other hand the Indirasagar dam is also situated on the same river in Khandwa district.

The Moghat reservoir of Khandwa is situated 3 kilometer away in the north east area of the city on 21° 49' 36" N latitude, 76°20' 56" E longitude and 324.4 meter, above from mean sea level. It is a man made reservoir built in 1897 which receives rainwater through two main sources one is called Ajanti canal and other is a local nalaha called barud nalaha, from its 23.30 sq. kilometers catchment's area. The storage area of Moghat reservoir is 2.02 sq. km., which has 5.36 kilometers shore line surrounded by hills with some large trees shrubs and agricultural land.

No known source of is there, except the rainwater, which brings agricultural waste and other organic matter from the catchments area of the reservoir. Besides this animal and human activities especially in summer season gives some sort of pollution to the reservoir.

Many researchers have done studies on physico-chemical and biological characteristics of river dam and lake water. David et al. (1969), George (1976), Sharma (1980), Trivedi and Goel (1990), Rao (1991), Khanna (1993), Sharma S.K. et al. (2003), Pathak et al. (2004), Nisar Sheikh and Yeragi (2004). But the informations are either lacking or insufficient from the limnological aspect of several reservoir systems of India. The importance of morphometric and hydrobiological details of reservoirs to assess the physico-chemical properties is being realized in the work of Symon et al. (1965), Neel (1966) and Fee ((1976). The importance of Indian reservoirs has been pointed out by Sreenivasan (1969), Jhingran (1982) and Rao (1988). In the state of M.P. biological study work made by Rao in narmada river system Khare and Unni (1986) on Kolar river Rao and Diwan (1979), Rao and Jain (1985) on Chambal and Khan river.

MATERIAL AND METHODS

The water sample were collected monthly from four sampling stations for physico chemical analysis through out the study period of one year (Oct. 04 to Sept. 05). The samples are well mixed and stored in two litter plastic cane. Sample collection was usually completed during morning hours between 7.00 to 9.00 a.m. The water temperature, pH and dissolved oxygen were estimated on the spot at the time of sampling while other parameters were estimated in the laboratory, standard methods as given by Trivedi and Goel (1986), Saxena (1990) and APHA 1992) were followed for estimation of various physicochemical parameters of water.

RESULTS AND DISCUSSION (see table 10.1)

Water temperature : The temperature of water were found in the range between 17.5 to 29.8. The temperature of water was maximum in the month of may and minimum in the month of December. The temperature is one of the important physical parameter which directly influence some chemical reaction and the life of the animals in water. Air temperature and water temperature are directly correlate with each other as recorded by many workers. Shinde (1995) recorded the temperature range of Tanas river from 20 to 32°C. The oxygen value goes down with high temperature.

Turbidity : In present study the turbidity of water were recorded maximum (133.5 NTU) in the month of July and minimum (19.2 NTU) in November. The turbidity of any water body is effected by the factors like planktonic growth, rain fall, water level and weather condition. The higher value of turbidity in July could be due to zooplanktonic saturation of summer and rain water inters from catchment area which has direct correlation with turbidity.

Dissolved solids: The dissolved solids have been reported important for the productivity of aquatic environment. In present study the minimum value of dissolved solids were observed 102.8 mg/lit. in February and maximum 210.8 mg/lit. in July. Mishra and Saxena (1993) reported dissolved solids range of 89 to 296 mg/lit. at Goury tank Bhind. Naik and

Table: The mean value of various parameters of four sampling Table:

	Oct.-04	*Nov.*	*Dec.*	*Jan.-05*	*Feb.*	*March*	*April*	*May*	*June*	*July*	*Aug.*	*Sept.*
Temp.	29.3	21.05	17.5	20.4	22.4	25.8	27.6	29.8	29	29.7	26.8	28.3
Turbidity	20.4	19.2	22.24	30.25	32.5	50.25	81.75	83.5	104.25	133.5	117.5	37.25
D Solids	175.75	168.2	159.3	124.25	102.8	130.2	151.5	189.5	210.3	210.8	187.7	183.2
PH	8.25	8.23	8.24	8.21	8.22	8.26	8.28	8.32	8.28	7.76	7.72	8.22
Alkalinity	124.4	121.8	118.4	114.2	117.1	119.4	121.7	125.5	127.3	100.7	99.1	122.8
D Oxygen	5.9	8.4	8.9	8.1	7.84	6.9	5.27	4.9	5.1	5.5	5.3	5.9
THard	124.5	119.5	120.45	121.05	122.75	126.25	134.3	144.7	152.5	102.2	94.8	93.1
Ca Hard	29.23	27.73	27.75	28.45	28.98	30.28	32.63	35.68	36.94	28.98	27.8	27.6
Mg Hard	12.28	13.1	14.33	14.72	15.78	16.74	17.55	24.6	25.9	10.62	10.44	19.4

Purohit (2001) reported total dissolved solids range 151 to 238 mg/lit at river Brahmini. The present investigations support the studies conducted by above mentioned researchers.

pH : The water of Moghat reservoir is alkaline and ranged from 7.72 in August to 8.32 in May the low value of pH in August is due to rain water and the higher value in May is directly related to high water temperature and low level of water. Das et al. (1961) observed that high pH value coincided with plankton peak Anand and Sharma found the pH values between 6.85 to 8.51 in Mansar lake. Shaikh Nisar and Yeragi recorded pH range between 7.04 to 8.43 in Tansa river Thane district Maharashtra.

Alkalinity : The total alkalinity ranged from 99.13 in August to 125.5 in May the alkalinity might be due to high pH. The high pH may be due to the hydroxide, carbonates and bicarbonates. These substances should be present in water to support growth of aquatic organisms. Rupwati and Radhakrishnan (1983) in the study of quarry pools observed low alkalinity during monsoon. Chawan et al. (2004) found the alkalinity values varied from 128 to 261 mg/lit. in Manjara project reservoir in district Beer Maharashtra. Low value of alkalinity in August might be due to dilution of carbonates and bicarbonates by rain water.

Dissolved Oxygen : D.O. in Moghat reservoir varied between 4.9 p.p.m. in May to 8.9 ppm in December. D.O. concentration remained low during April to October and higher from November to March. Winter rise in D.O. has also earlier been recorded by Singh et al. (1980) and Rao (1986) and it appears to be due to its greater solubility, reduced microbial decomposition of dead organic matter and low demand of organisms for respiration at low temperature. Low value of D.O. during April to October due to high temperature and dead organic matter brought in to reservoir from the catchment area during early monsoon. Increased organismal respiratory demand at high temperature is also support this low D.O. value in summer Rao (1986).

Calcium, Magnesium and Total Hardness : The present study recorded the mean values of Ca, Mg and Total hardness as 27.08 mg/lit. in August to 36.95 gm/lit. in June, 11.9 mg/lit. in Sept. to 25.9 mg/lit. in June and 93.1 mg/lit. in Sept. to 152 mg/lit. in June respectively. The low values of all three, estimated in monsoon (July, Aug., Sept.) and maximum values were recorded in summer (June) similar results were observed by Koshy and Nayer (1999) Mishra and Tripathi (2001) recorded high values during summer and low values in monsoon in Ganga river. Ca and Mg both produces hardness to the water and form most abundant ions in the fresh water. The average values of calcium, magnesium and total hardness never exceeded the standard limits of W.H.O. (1984) it means that the water of Moghat reservoir is not hazardous to the animals and human beings who are using it for their daily needs but it should not be taken by human beings without proper treatment for purification.

ACKNOWLEDGEMENT

Authors are thankful to the principal Govt. Girls P.G. College Motitabela Indore for providing Lab. facilities for analyzing the different parameters.

REFERENCES

APHA (1992) Standard methods for examination of water and waste water.

Belsare et al. (1990) Limnological studies on Bhopal lake seasonal change in a biotic factors in the upper lake Proc. Nat. Acad. Sci. 60:223-227.

Das S.M. (1961) Hydrogen concentration and fish in fresh water eutrophic lakes of India Nature 20 London.

Ganpati S.V.(1943) Seasonal changes in the physical and chemical conditions of garden pond containing abundant aquatic vegetation J. Madras univ. 13, 55-69.

George J.P. (1976) Hydro biological studies on lower lake Bhopal with special reference to the productivity of economic fishes. Ph.D. thesis Bhopal Univ. Bhopal.

Joshi B.D. et al. (1993) Some aspects of physico-chemical characteristics of water of Ganga canal near Jwalapur, Hardwar. Himalayan J.Envi. 7(1) 76-82.

Mishra and Tripathi (2001) Impact of city sewage discharge on physico-chemical condition of Ganga water Asian J. of microbial biotech. and Env. Sci. Vol. 930, no.(4) 333-338.

N. Subhash Chandra Meitei, P.M. Patil and Bhosle A.B. (2004) Physico-chemical analysis of Purna river potability J. aqua Biol. Vol. 19(1) pp. 103-105.

Patel Pushpa, Mahajan S.K. and Pathak S.K. (2002) Hydrobiological observations on viral reservoir/Khargone M.P. Enviorn. Cone. J. 3(2):57-59.

Rao K.J. (1986) Studies on the seasonal and diel. Variations in some physico-chemical conditions of fresh water under prawn culture Proc. Nat. Symp. Fish and Env. 96-102.

Singh R.K., Shrivastava N.P. and Desai V.R. (1980) Seasonal and diurnal variations in physico-chemical conditions of water and plankton in lotic sector of Rihand reservoir U.P. J. Inland Fish India 12(1): 100-111.

Saxena (1990) Environmental analysis water, soil and air Agrobotenical Publishers India 184 pp.

Sharma Anjana S.K. Sharma and L.K. Mudgal (2003) Indore. J.com. toxicology and physiology 1: 22-25.

Trivedi R.K. and P.K. Goel (1986) Chemical and biological methods for water pollution studies Environmental publication Karad.

11

Haematological Manifestation in Relation to Ecophysiological Adaptations in a Freshwater Air-breathing Teleost, *Channa gachua* (Ham.)—Hormonal Regulation

Abhiranjan Kumar and B.N. Pandey

SUMMARY

Effect of different hormones and pharmacological drugs on certain haematological parameters (viz, RBC counts, Hb content, PCV, MCV, MCHC, cell size, surface area of erythrocytes and oxygen capacity etc.) were made in a fresh water air breathing snake headed murrel teleostean fish, *Channa guchua* (Ham.). The treatment of L-thyroxine (T_4), T_3, methyltestosterone, hydrocortiosone, ACTH, adrenaline, phentolamine and Growth Hormone (GH) brought significant ($p<0.05$) increase in RBC counts, Hb content, PCV and oxygen capacity as compared to control group which were subjected to the treatment of seasame oil. The MCV, MCHC and SAE showed decreasing trend in these experimental animals. Though the treatment of thiouracil failed to alter changes in such parameters to a significant level but a tendency towards increase of RBC count, Hb content and PCV values was observed. Possible reason and mechanism of such changes in different blood parameters in relation to ecophysiological adaptations have been discussed in this paper.

Dept. of Zoology, Magadh University, Bodh Gaya - 824234, Bihar, India.

Key words: *Channa gachua,* haematology, hormones, Ecophysiology.

INTRODUCTION

Organisms live in a particular environment which is subject to marked seasonal fluctuations (Wolfson, 1964). Seasonal changes in the environmental conditions viz, temperature, day length (photoperiod), rainfall, food and physico-chemical composition of the medium have pronounced effect on the life of poikilothermic animals. Successful living demands a continual adjustment or regulation of the internal media with the changing environmental condition and the endocrine system is intimately involved in their control (Hoar. 1959). The thyroid, gonads and interrenal tissues through their action on development, growth, metabolic rate and reproduction help the ectothermic animals to adapt successfully to change in their external environment.

Relationship between seasonal changes in ambient water temperature, day length, rain fall vs. blood parameters and activity of endocrine glands has already been established in *Macrognathus aculeatum* (Prasad *et al.,* 1978).

Since the blood if the carrier of respiratory gases and are greatly under the influence of various environmental and endocrine factors, the study of blood parameters in relation to hormonal treatment becomes important and inevitable.

The present work is an endeavour to study the effect of different hormones and pharmacological drugs on some haematological indices (such as RBC counts, Hb content, PCV, MCH, MCHC, cell size and oxygen capacity etc.) in this fish as this aspect could not attract proper attention of Ichthyologist and endocrinologist in spite of the fact that application of haematological indices and criteria for diagnosis of the state of health and diseases of fish and assessment of pollution and its effects has been duly emphasized in the past (Mawdesley-Thomas, 1971; Prasad et al. 1978).

MATERIAL AND METHODS

Channa gachua, an air breathing snake headed murrel teleostean fish is found in fresh waters throughout the greater part of India and in Andaman Islands, Ceylon and Burma; body brown above with eight dark bands running forwards from dorsal ridge to just below lateral line; length varies from 16-33.0 cm. It builds its nest in sheltered crevices in the bank.

Live specimens of *Channa gachua* ware collected from Dighi and Bisar tank of Gaya and transported to the laboratory in a plastic container. In the laboratory the fishes were treated with potassium permanganate for few minutes and then transferred to glass aquaria. The fishes were kept for seven days with proper diet so as to acclimatize them in the laboratory condition. Only adult fishes of approximately the similar body weight (45.0$\pm$1.5 g) were used in the present study.

Fishes were divided into ten groups (1-10) each comprising six individuals. Group I got the treatment of seasame oil (Sham-treated) while rest of the fishes (2-10 groups) got the treatment of different hormones (dosage prepared with seasame oil as carrier solution). Different hormones (as required; Table 11.1) were injected intraperitoneally to the fish on ventral side just in front of pelvic fin. Before any experiment the effective dosage of different hormones were standardised and the blood was taken out for haematological studies six hours after the last injection.

The details of the methods used in the study of different blood parameters were those of Pandey et al., 1997; Singh, 1997 and Waquas, 2002. They used Neubauer double haemato-cytometer for RBC counts and Sahli's haematometer for the determination of Hb content. PCV (packed cell volume) was determined with the help of microhaematocrit pipette after Van Allen. The haematocrit tube was centrifused for 30 minutes at 4000 RPM. The oxygen capacity of the blood was calculated by multiplying the total Hb content with the oxygen combining power of 1.25. ml of oxygen per gram of Hb (Johansen, 1970). It was expressed as volume percent (vol. %).

The difference of significance, if any, between the control and experimental animals, was calculated by student's t-test at the level of 5%.

RESULTS

The data showing the effect of different hormones and pharmacological drugs on RBC counts, Hb content, PCV, MCV, MCHC, cell size, surface area of erythrocytes (SAE) and oxygen capacity in control and experimental group of fishes have been summerised in Table 11.1. A perusal of this table indicates that:

1. In control group the RBC counts, Hb content, PCV, MCV, MCHC, cell size, SAE and oxygen capacity values were - 2.26 million/mm^3, 13.5 g%, 35.5%, 157.77 μm^3, 38.02%, 9.142 x 8.341 μm, 78.50 μm^2 and 16.87 vol% respectively.
2. The treatment of L-thyroxine (T_4), (T_3), methyltestosterone, hydrocortisone, ACTH, adrenaline, phentolamine and GH brought significant ($P<0.05$) increase in RBC count, Hb, PCV and oxygen capacity as compared to control while thiouracil failed to alter the above noted parameters to a significant level.

DISCUSSION

Hormones play on important role in eco-physiological adaptations in poikilotherms. For ecological adaptations it is not animals which as whole change but they gradually acclimatize and adjust their physiological machinery to the changing environmental conditions. In physiological adaptations, respiration and blood is supposed to be the most important aspects of a fish, hence the studies on the effect of hormones on fish blood becomes essential.

Our knowledge with regards to the physiological role of hormones in poikilothermic animals and specially fishes, is very limited. It is, therefore, inevitable to make such investigation in Indian teleostean fishes. The values obtained for RBC counts, Hb content and PCV etc. are very close to that reported in a number of Indian air breathing fishes by several investigators (Pandey et. al., 1976, 1978; Dubey and Munshi, 1973; Singh, 1997 and Waquas, 2002).

In the present investigation in *Channa gachua,* the treatment of phentolamine and hydrocortisone brought significant

Table 11.1: Effect of some hormones, L-thyroxine, T_3. methyl testosterone, growth hormone, hydrocortisone, ACTH and adrenaline on haematological parameters in *Channa gachua* (Ham). Water temp. = 27.0°C ± 1.0°C; = = sd; N = 6 for each group; SAE = Surface Area of erythrocyte; Body weight 45.0 ± 1.5

Sl. No.	Condition	Dose (mg/ 100g)	Injec -tion in no. of days	RBC counts (10^6/mm^3)	Hb content (g %)	PVC (%)	MCV (μ^3)	M.C.H .C. %	Cell size (µm)	SAE (µm^2)	Oxygen capacity (Vol.%)
1.	Control	-	-	2.26 ± 0.11*	13.5 ± 0.45	35.5 ± 0.34	157.77	38.02	9.142 x 8.341	78.50	16.87
2.	L-Thyroxine (T_4)	0.5	2	2.62 ± 0.08*	13.9 ± 0.10	36.7 ± 0.31	147.52	38.09	9.312 x 8.286	77.16	17.37
3.	T_3	1.0	4	2.48 ± 0.11*	13.8 ± 0.13	35.7 ± 0,25	143.95	38.65	9.316 x 8.290	77.23	17.25
4.	Thiouracil	2.0	8	2.35 ± 0.19^{NS}	13.6 ± 0.12	35.5 ± 0.23	151.06	37.75	9.412 x 8.336	78.46	17.00
5.	Methyltestosteron e	0.5	2	3.40 ± 0.13*	16.1 ± 0.11	39.2 ± 0.00	115.29	41.07	-	-	20.12
6.	Hydrocortisone	1.0	4	4.12 ± 0.17*	19.5 ± 0.16	54.2 ± 0.45	131.55	35.98	8.812 x 8.011	70.59	24.37
7.	ACTH	4 IU	4	2.74 ± 0.11*	15.3 ± 0.16	41.2 ± 0.41	150.36	37.13	9.221 x 8.154	75.19	19.12
8.	Adrenaline	0.12	1	3.42 ± 0.13*	17.4 ± 0.18	48.4 ± 0.28	141.52	35.95	9.212 x 8.221	75.73	21.75
9.	Phentolamine	1.0	4	3.84 ± 0.15*	19.2 ± 0.19	52.3 ± 0.65	150.28	36.71	9.012 x 8.150	73.45	24.00
10.	Growth hormone (GH) or Grorm inj	1.5	5	3.31 ± 0.12*	17.3 ± 0.11	48.0 ± 0.38	145.01	36.04	9.201 x 8.156	75.04	21.62

increase in RBC count, Hb content, PCV value and oxygen capacity, a view consistent with the finding of Singh (1997).

Chatterjee (1975) is of the opinion that increase in RBC count and Hb content after hydrocortisone treatment may be due to increased loss of water by the kidney and passage of more water from blood streams into the tissue spaces. Munshi et al. (1977) have reported that the treatment of hydrocortisone in *H. fossilis* leads to decrease in blood volume. Thus, the dehydration of blood leading to an increase in different blood constituents seem to be logical.

Injection of adrenaline and ACTH brought significant increase in RBC counts, Hb content and PCV values, as compared to control group of *C. gachua.* The increase in such blood parameters after such treatment is usually believed to be due to mobilisation from the blood depot, specially due to contraction of spleen (Best and Taylor, 1967; Chatterjee. 1975). The ability of increasing rapidly the number of erythrocytes in the blood by increasing those stored in spleen (Toft, 1975) seem to be an important factor in the adaptability of the organism to frequent changes in external or internal environment..

The information regarding the effect of thyroid gland on haematological parameters in vertebrates is very limited (Adeniyi and Ononoze, 1990). Injection of T_3 and L-thyroxine (T_4) resulted in an increase in RBC counts, Hb content, PCV values and oxygen capacity. Treatment of thyroxine influences protein metabolism and the electrolyte balance which is seen in the deposition of intracellular mucoprotein accompanied by an increase in the intracellular level of salt and water and modification in osmotic balance is further reflected in the blood by an increased protein content and reduced volume (Zarrow et al 1964). Gabos et al. (1973) in *Cyprinus carpio,* Munshi et al. (1977) in *H. fossilis,* Pandey et.al. (1978) in *C. batrachus* and Singh (1997) in *Mystus vittatus* have reported that thyroxine is involved in the elimination of water from the body which is reflected in decrease in blood volume in these species (haemoconcentration), ultimately resulting in an increase in RBC counts and Hb content. Thiouracil failed to alter blood parameters to a significant level in *C. gachua* though a tendency of increase in such parameters was noticed, a view consistent with the finding of Singh (1997).

Treatment of methyltestosterone and GH brought an increase in RBC counts, Hb content, PCV and oxygen capacity in *C. gachua* in the present investigation. Guyton (1971) has stated that when normal quantities of testosterone are injected into a castrated adult (Man), the increase in RBC count/mm^3 of blood is approximately 20%. Though the exact reason of increase in RBC count could not be understood but this difference may be partly due to the increased metabolic rate following testosterone and GH administration rather than to a direct effect of these hormones on RBC production. Hoar (1958) in gold fish and Pandey (1976) in *H. fossilis* have reported an increase in oxygen consumption after the injection of testosterone in these species. Thus, the increase in different blood parameters, after the injection of methyltestosterone in *C. gachua* seems to be quite meaningful. Ratan (1975) has reported that erythropoiesis is stimulated in man after testosterone treatment.

It can't be said with conformity that the increase or decrease in various blood components after the treatment of different hormones and pharmacological drugs in this fish are due to increased erythropoetic intensity or haemolysis respectively. But it seems more probable that the different hormones and pharmacological drugs used in the present investigation are in one way or other associated with the maintenance of ionic balance between cardiovascular system and intracellular space which ultimately influence total blood volume. Thus, the increase (Haemoconcentration) or decrease (Haemodilution) in various blood parameters may be because of the increase or decrease in blood or plasma volume respectively. It requires further investigation.

ACKNOWLEDGEMENT

Financial assistance from C.S.I.R., New Delhi to one of us (A.K.) is greatly acknowledged.

REFERENCES

Adeniyi, K.O. and Ononoze, M.E. 1990: The effect of thyroidectomy on total and differential leucocyte counts in Rat. *Him J. Env.* Zool. **4(2)**: 140-143.

Best, G.H. and Taylor N.B. 1969: *The physiological basis of medical practice,* Baltmore William and Wilkins company. Scientific book Agency, Kolkata.

Chatterjee, C.C. 1975: *Human Physiology.* Vol. 1 Medical Allied Agency. Calutta, pp. 1-764.

Dube, S. C., Munshi, J.S.D. 1973 : Quantitative study of erythrocytes and haemoglobin in the blood of an air-breathing fish, *Anabastestudineus* (Bloch) in relation to its body size,. *Folia haematol,* Leipzig. **100(4)**: 436-446

Gabos, M; C.C. Pora, E.A. and Race, L. 1973: Effect of the L-thyroxine (T_4), TSH and thiourea (Tu) treatment on the oxygen consumption of the carp. *Stud. Cercet. Biol.* Zool. **25 (1)**: 39-43.

Guyton A.C. 1971 : *Text book of medical physiology,* 4th edition WB Saunders Company. Philadelphia London Torento.

Hoar, W.S. 1958: Effect of synthetic thyroxine and gonadal steroid on the metabolism of goldfish, *Can. J.* Zool. **36:***113-121.*

Hoar, W.S. 1959. In *"Comparative Endocrinology"* Eds. A. Gorbman. John Wiley and Sons. New York: 1-23.

Johansen, K. 1970: Air breathing in fishes. In "fish Physiology", vol. IV eds. W.S. Hoar and D.J. Randall. Academic press. New York and London, pp. 367-411.

Mawdesley - Thomas, Le 1971 : Toxic chemicals, the risk of fish. *New Scientist,* **49**:74-75.

Pandey. B.N. 1976: Effect of gonadal steroids on oxygen consumption in *Heteropneustesfossilis(Bloch), Pol. Arch. Hydrobiol,* **123**:155-160.

Pandey, B. N.; Pandey; P.K. Choubey, B.J. & Munshi, J.S. D. 1976 : Studies on blood components of an air breathing siluriod fish, *Heteropneustes fossilis* in relation to body weight. *Folia haematol,* **103**.101-116.

Pandey B. N.: Chanchal, A. K.; S.B. Singh 1978 : Effect of hormones and pharmacologica drugs on the blood of *Anabas testudineus* (Bloch) *Folia haematol.* **105** (5): 665-651.

Prasad, S.; Pandey, B.N. Sinha, D.P. and Chanchal, A.K. 1978: Haematological studies on the Indian freshwater mul-eel,

Macrognathus aculeatum, Z. TierPhysiol and Futter mittes - kumdle. B and **40** Heft. 6, 5 :295-301.

Ratan, V. 1975: Hand book of Human Physiology. Jaypee Brothers. New Delhi.

Singh, Kumari, Saphalta 1997 : Role of hormones in the regulation of haematological parameters in the catfish, *Mystus vittatus* (Bloch). Ph.D. Thesis. Magadh University, Bodh Gaya.

Toft, W. 1975: Under die Blutbildung and Blutbildungsstaffen beim karpfen *(Cyprinuscarpio). Z. Fischerei* **4(3-4)**: 257-288.

Zarrow, M.X.: Yochim, H.M.; McCarthy, J. L. and Sanborn, R.C. 1964: Experimental Endocrionology. Academic Press. Inc. New York, London.

Waquas, S. (2002): A Comparative study of haematology of Indian obligate and facultative air breathing fishes. Ph.D. Thesis. Magadh University, Bodh Gaya.

Wolfson, A. 1964: Animal photoperidism. "In photophysiology." Eds. A.G. Giese. Academic Press. New York. 47-58.

Study of Haematological Parameters in the Fresh Water Fish *C. Catla* (Ham.) After Exposure to the Alcoholic Extract of Fruits of *A. Sinuata* (Dec.) Merr.

*R.G. Patil and S.G. Nanaware**

SUMMARY

Fresh Water fish *C. catla* was exposed to the sublethal concentration of alcoholic extract of fruits of *A. sinuata* in, laboratory conditions for 120 hrs. The piscicidal compounds in the fruits of *A. sinauta* induced the haemolysis in fresh water fish *C. catla* with increased period of exposure maximum decrease in the counts of Leucocytes (8:94%), erythrocytes (5.21%), Hb (12.37%) and haematocrit (8.42%) were occurred after 120 hrs. While Erythrocyte sedimentation rate was increased with the increased exposure period. Since plant toxin induced haemolysis decreases the oxygen carrying capacity of blood which further affects the metabolic activities in the fish *C. catla*. The results are discussed in relation to the mortality rate of this commercially important fish in the pisciculture.

Key words: Piscicidal Plants, Fish, Haematology.

INTRODUCTION

The haematological tests give significance in assessing the physiological responses of fishes (Joshi et al., 1980). These tests are also used to diagnose disease caused by various factors viz. environmental stress, starvation, toxicants, infections, etc. in

---

Head, Department of Zoology L.B.S. College, Satara - 415 002.

* Department of Zoology Shivaji University, Kolhapur - 416 004. Maharashtra (India)

pathological laboratories. But it is not documented in piokilotherms especially in fishes (Wedmeyer and Yasutake, 1977).

Recently haematological experiments have been done to measure the impact of various toxicants (Agrawal et. al, 1978; Buckley, 1976; Joshi, 1978; Newman and Mclean, 1974 and Mcleay and Gordon, 1977). However, the studies on the effects of piscicidal compounds of plants on fish are meagre (Bhatt and Singh, 1985). Hence, the present paper reports, the effect of plant toxin from fruits of Acacia sinuata on haematological parameters of fresh water fish Catla catla.

MATERIAL AND METHOD

The fruits of *A. Sinuata* were collected, air dried and powdered mechanically. This powder was then extracted in absolute ethanol. The ethanol extract of fruits of *A. Sinuata* wad dried in vacuum desicator.

The healthy adults with average length 12 cms. and average weight 150 gms. were collected from the local tanks. Then fishes were kept in glass acquaria with continueous supply of tap water. After fascilitating acclimatization in laboratory conditions for a week, ten fishes were exposed to the sublethal concentration (110 ppm) of ethanol extract of fruits of A. Sinuata, a control set was also maintained. At the interval of 24, 48, 72, 96 and 120 hours ol intoxication two fishes were taken out and anaesthetized (Tricaine methane sulphonate) and dried with filter paper. Blood sample was collected in syringe rinsed with anticoagulent, by restoring to cardiac puncture. Blood was stored in a glass tubes layered with anticoagulant (EDTA).

The total Red blood cell count and White blood cell count was determined through improved Neubaur's haemocytometer (Germany) by using Hayem's diluting fluid and 30% glacial acetic acid with methyl voilet respectively. Haemoglobin concentration of blood was determined by Sahli's haemometer (Germany). Haematocrit values (Packed cell volume (PCV) from blood was determined with Wintrobe's haemotocrit pipettes, for this purose blood was centrifuged for 30 minutes at 4000 rpm. The Mean corpuscular volume (MCV), Mean cell haemoglobin (MCH) and Mean corpuscular haemoglobin concentration (MCHC) was calculated by standard methods (Bauer, 1990). Erythrocyte sedimentation rate (ESR) was determined by Westergern method (Bauer, 1990).

OBSERVATIONS

Sublethal concentration of ethanol extract of fruits *A. Sinuata* showed haemolytic effects in fresh water fish *C. catla* from 12 hours to 120 hours.

Table 12.1 shows values of R.B.C. count, W.B.C. count, Haemoglobin percentage, haematocrit, Mean Cell Volume (MCV), Mean cell haemoglobin (MCH), and Mean cell haemoglobin concentration (MCHC). All these values are observed to be decreased with increased period of intoxication. It also shows values of erythrocyte sedimentation rate (E.S.R.) which was found to be increased with increased period of intoxication.

Table 12.1 also shows the percentage decrease and increase in the different haematological parameters. Minimum percentage decrease in R.B.C. count, Haemoglobin percentage, PCV and MCV was 0.99%, 1.03%, 5.01%, 0.46%, 1.03%, 1.02%, 0.03%, Nil at 24 hours respectively. While maximum decrease in R.B.C. and W.B.C. count, Haemoglobin percentage, PCV, MCV, MCH and MCHC was 5.21%, 8.94%, 12.37%, 8.42%, 5.54%, 7.52% and 4.32% at 120 hours respectively.

Erythrocyte sedimentation rate (E.S.R.) was found to be increased. Minimum increase was 2.12% at 24 hours while maximum increase in E.S.R. was 15.15% at 120 hours.

RESULTS AND DISCUSSION

Table 12.1 presents data on the effect of piscicide from alcoholic extract of fruits of *A. Sinuata* on different haematological values in treated fish are compared with the control set, it is observed that values of R.B.C. and W.B.C. count, Haemoglobin percentage, PCV, MCV, MCH and MCHC are decreased. Maximum percentage fall in R.B.C. and W.B.C. count Haemoglobin percentage, PCV, MCV, MCH and MCHC was

Table 12.1: Effect of Plant Toxin A. Sinuata on Haematological Parameters of Fish C. catla.

No	Haematologcial Parameters	Control	Intoxication Period (In hrs.)									
			24	%	48	%	72	%	96	%	120	%
1	R.B.C.× 10^6 × mm^3	4.03	3.99	0.99	3.96	1.74	3.91	2.98	3.89	3.47	3.82	5.21
2	Hb (gm %)	9.70	9.60	1.03	9.50	2.06	9.30	4.12	8.70	10.31	8.5	12.37
3	PCV (%)	39.20	38.80	1.02	38.60	1.53	37.20	5.10	36.70	6.38	35.9	8.42
4	MCV(um)	97.27	97.24	0.03	97.47	(0.21)	95.15	2.18	94.34	3.01	91.88	5.54
5	MCH (pg)	24.06	24.06	- -	23.98	0.33	23.98	0.33	22.36	7.07	22.25	7.52
6	MCHC (%)	24.74	24.74	-	24.61	0.53	25.00	(1.05)	23.70	4.20	23.67	4.32
7	ESR (mm/hr)	3.30	3.37	(2.12)	3.40	(3.03)	3.55	(7.58)	3.75	(13.64)	3.80	(15.15)
8	RBC-C:N	3.20	3.40	(6.25)	3.44	(7.50)	3.84	(20.0)	3.81	(19.06)	3.81	(19.06)
9	RBC-N:C	0.31	0.30	3.23	0.29	6.45	0.25	19.35	0.263	15.16	0.26	15.16
10	Clotting time (sec.)	29.30	28.70	2.05	28.30	3.41	. 27.70	5.46	26.90	8.19	26.10	10.92
11	W.B.C.× 10^3/mm^3	35.80	36.40	(1.68)	36.00	(0.56)	34.70	3.07	33.40	6.70	32.60	8.94
12	THC.× 103/mm^3	41.10	41.80	(1.70)	42.2	(2.68)	42.90	(4.38)	43.10	(4.87)	43.90	(6.81))

NB: The percentage (%) values shown in the bracket are positive (+ve), while others without brackets are negative (-ve)

8.94%, 5.21%, 12.37%, 8.42%, 5.54%, 7.52% and 4.32% respectively at 120 hours of intoxication. All these values decreased with increased period of intoxication.

The fall in the data of R.B.C. and W.B.C. count, Haemohglobin percentage, PCV, MCV, MCH and MCHC in C. catla after intoxication with sub-lethal concentration of piscicide has been in agreement with Pandey et al, (1976), Hooper and Sunderman (1978), Joshi et al., (1979), Agrawal et al. (1979) who have reported among malthion, folidal, nickel, zinc and cobalt treated fishes.

In present study when fishes are treated with the piscicide, it enters into the body and then into the blood streams of fishes. The entry of piscicide into the body of C. *catla* induces lysis of erythrocytes and leucocytes. Injury to the blood cells causes decreased R.B.C. and W.B.C. count, Haemoglobin percentage, PCV, MCV, MCH and MCHC. This type of fall in haemolytic parameters due to toxicants were demonstrated by Kptoom et al., (1982) and Bhatt and Farswan (1987). From this discussion the pescicide appears to be haemolytic in nature.

Decreased PCV and increased Erythrocyte sedimentation rate (ESR) is due to the degradation of blood proteins (Bhatt, 1985) in the intoxificated fish.

Hence it is concluded that plant toxin induced haemolysis affects the oxygen carrying capacity of blood which further affects the metabolic activities in the fish.

ACKNOWLEDGEMENTS

We are thankful to Prin. Abhayakumar Salunkhe, Karyadhyaksh, Swami Vivekanand Shikshan Sanstha, Kolhapur, Prin. J.L. Ayanapure (M.L.C.), Secretary Swami Vivekanand Shikshan Sanstha, Kolhapur and Prin. Suhas Salunkhe, L.B.S. College, Satara for continueous encouragement and providing necessary facilities.

REFERENCES

Agarwal, N.K. and Mahajan, C.S. 1983 Haematological and haematopoietic studies in pyridoxine deficient fish, *Channa punctatus* (Bloch). J. Fish Biol. 22, 91-103.

Agrawal, S.J., A.K. Srivastava, and H.S. Chaudhry 1979. Haematological effects of nickel on a fresh water toleost *Colisa fasciatus.* Acta. Pharmacol. et Toxicol.45,215-217.

BauerJohn D. 1990. Clinical laboratory methods (pp. 178 to 191) Ninth edition, published by B.I. Publication, 54, Janapath, New Delhi-110 001.

Bhatt, J.P. and H.R. Singh 1985. Effects of Engelhardtia colebrookiana (Lin.) on a fresh water teleost *Barilius bendelisis* (Ham.). Science and Culture 51(A) 132-133.

Bhatt, J.P., A.K. Dobryal and Y.S. Farswan 1987. Growth response in the fry of *Schizothrorax richardsonii* (Gray) to the plant toxins. J. Environ. Biol. 8(2), 207-215.

Bhatt J.P. and Farswan Y.S. 1992. Haemolytic activity of Piscicidal Compounds of Some Plants to a Freshwater Fish *Barilius bendelisis* (Ham.) J. Environ Biol. 13 (4), 333-342.

Buckley, J.A., C.M. Whitemove and R.I. Matsuda 1976. Changes in blood chemistry and blood cell morphology in cohosalmon *(Onchorvnchus kisutch)* following exposure to sublethal level of total residual chlorine in municipal waste water J. Fish. Res. Bd. Canada, 33, 776-782.

Garg Veena, Tyagi S.D., Singh Neera and Agarwal S.C. 1991. Thallium Nitrate induced haemato- biochemical analysis of *Heterponeustes fossilis* blood, J. Environ, Biol, 12(3), 319-323.

Hopper, S.M. and F.W. Sunderman 1978. Maganese inhibition of nickel sulfide on erythrocytosis in rats. Res. Commun. Chem. Pathol. Pharmacol. 19,337-345.

Joshi, B.D. 1978 effects of asphyxiation on some haematological values of the fish *Clarias batrachus.* Proc. All India Sem. Ichthyol, 5-8.

Joshi, B.D. L.D. Chaturvedi and R. Dabral 1979. Changes in the blood cell components of *Heteropneustes fossilis* following transfer of Polidol water at low temperature Proc. Symp. Life and Taxic Environ, 86.

Joshi, B.D., Chaturvedi, L.D., and Daliral, R. 1980 Some haematologic values of Clarias batrachus following its sudden transfer to varying temperatures. Indian J. exp. Biol., 18,76-77.

Kiptoon, J.C., G.M. Mugera and P.G. Waiyaki 1982. Haematological and biochemical changes in cattle poisoned by *Gnidia latifolia* Sin. *Lasiosiphon latifolius* (Thymelaeaceae). Toxicol. 25,129-139.

McLeay, D.J. and M.R. Gordon 1977. Leucocrit: A simple hematological technique for measuring acute stress in salmonid fish. J. Fish. Res. Bd. Canad, 34(9), 2164-2175.

McLeay, D.J. 1973 Effect of a 12 h and 25 days exposure to kraft pulpmill effluent on the blood and tissue of juvenile coho salmon, J. Fish. Research Bd. Canada, 30.

Newmaii, M.W. and S.A. Mclean 1974 Physiological response of the cunner, Tautogolabrus adspersus to cadmium, 6, NMFS-S SRF 681, 27-30.

Pandey, B.N., A.K. Chanchal and M.P Singh 1976 Effect of malathion on oxygen consumption and blood of *Channa punctatus,* Ind. J. Zootomy 17(2), 95-100.

Thakur G.K. and Pandey P.K. 1990 BHC Poisoning Effect on Leucocytes of an Airbreathing Fish, *Clarias batrachus* (linn). J. Environ, Biol. 11(2), 105-110. 1991.

Wedemeyer, G.A. and W.T. Yasutake 1977 Clinical methods for the assessment of effects of environmental stress on fish health. US Dept. Interior. Fish and Wildl. Ser. Technical paper 89, 1-18.

13

Role of Nitrogen Oxides and Photo Chemical Smog in the Environmental Pollution

*Dilip Kumar Keshri**

SUMMARY

The Nitrogen oxides are most important gaseous air pollutions which are know to be injurious to the health of human and plants. These are produced primarily from combustion process of fuel (Petrol, Diesel) in automobiles and industries and secondarily from nitrogen fertilizer plants and burning of biomass. Nitrogen oxide and their secondary pollutants like pexoxyacetly nitrate or PAN and ozone are harmful to both mass and plants. The nitrogen dioxide (NO_2) is a pungent gas that produces a brownish haze and cause nose and eye irritation and pulmonary discomfort. The nitrous oxides decrease gaseous exchange in blood and hinder the functioning of lung NO_2 inhalation causes eye irritation, lung ocdeura, blood congestion. It is also nitrogen. The nitrogen dioxide, ozone and PAN severely injure many forms of plant life destroying the cells of leaves, damaging the chloroplarks, and interfering the with the plants metabolic processes. In bright sun light, Nitrogen oxides, hydrocarbons and oxygen interact chemically to produce powerful oxidant like ozone and peroxyacetyl nitrate called photochemical smog.

INTRODUCTION

Industrialization and urbanisation has resulted in a profound deterioration of air quality. There are seven primary

* Dept. of Zoology, T.S. College, Hisua, Nawada (Bihar)

air pollutants which are considered as health hazards when their concentration exceeds the national ambiant air quality standards. These are *suspended particulate matter, carbon monooxide, nitrogens dioxide, ozone and other photochemical oxidants (smog) lead and sulphur.* Among them, Nitrogen oxides are one of the common air pollutant which are known to be injurious to human health as well as plants. Nitrogen oxides include **NO, N_2O, NO_2, N_2O_4, N_2O_5** etc. The most common is nitrogen dioxide, which is a pungent gas. It produces a brownish haze, causes nose and eye irritation and pulmonary discomfort. **Statue of Liberty** is corroded from SO_2 and NO_2. Acid Rain is 60 - 70% due to SO_2 & SO_3 but 30 - 40% due to NO_2 and NO_3 Rainfall pH value has been estimated to be in between 5.5 to 6.9 in Bangalore, Delhi, Nagpur and Pune.

In the atmosphere, NO_2 is reduced by ultraviolet light to nitrogen monoxide (NO) and atomic oxygen (O).

$$NO_2 \longrightarrow NO + O$$

Atomic oxygen reacts with oxygen to form ozone (O_3)

$$O_2 + O \longrightarrow O_3$$

Ozone reacts with NO to form nitrogen dioxide (NO_2) and oxygen, thus closing the cycle.

$$No + O_3 \longrightarrow NO_2 + O_3$$

Sometimes, in presence of sunlight, atomic oxygen from the photochemical reaction of NO_2 also react with a number of hydrocarbons (such as methane, ethane, toluene etc. all of which originate from burning of fossil fuel or directly from plants) to form reactive intermediates called radicals. These redicals then take part in a series of reactions to form still more radicals that combine with oxygen, hydrocarbons, and NO_2. As a result nitrogen dioxide is regenerated. Ozone accumulates and a number of secondary pollutants are formed such as *formaldehyde, aldehyde* and *peroxyacetyle nitrate* or PAN ($C_2H_3O_5N$). All of these collectively form **photochemical smog.** Photochemical smog was first observed in Los Angeles and is related to smoke and Fog. In 1952, *London Smog disaster is* said to have caused the death of 4000 people.

ROLE OF NITROGEN DIOXIDE AS POLLUTANT

- The World Health Organisation reports about NO_2 that normal healthy humans exposed at rest or light exercise for less than 2 hours to concentrate above 4700 microgram/m^3 (2.5 ppm) experience pronounced derease in pulmonary function. The nitrogen oxides inhibit cilia action so that soot and dust penetrate far into the lung.
- Nitrogen oxides decreases gaseous exchange in blood and hinder the funtioning of lung. No_2 inhalation causes eye irritation, lung oedema, blood congestion. It is also a mutagen.
- Long term exposure to NO_2 at level upto 1 pm can cause both reversible and irreversible effects in the lungs resulting in emphysema like effects. Other effects have been observed in the spleen, liver and in blood. It causes irritation of nose and eyes.
- Oxides of nitrogen cause lesion, necrosis, defoliation, die back and death of many plants.
- When moisture accumulates in polluted air, the oxides of nitrogen form weak nitric acid, which are corresive to metal, stone, paint, rubber, textiles and even some plastics.

ROLE OF OZONE AND PAN AS POLLUTANTS

- Lower Concentration of ozone irritates the nose and throat while higher concentration of it cause dryness of the throat, headaches and difficulty in breating.
- Smog and air pollution are found to specifically damage pine forests, truck gardens crops, citrus groves, onion and celery fiedls and field crop of alfa-alfa and sweet corn.
- PAN impair photosynthesis, respiration, enzyme activity and amobolic processes of plants. These block the Hill reaction of photosynthesis.
- Ozone, PAN and NO_2 severely injure many forms of plant life, destroying the cells of leaves, damaging the chloroplast and interfering with the plants metabolic processes.

CONCLUSION

India's most severe environmental problem arises in several forms, including vehicular emissions and untreated industrial smoke. The rapid industrialization and urbanization has led to emergence of industrial centres without a corresponding growth in civic amenities and pollution control mechanisms. Efforts are currently under way to change this as new specification are being adopted for auto emission, which currently account for about 70% of air pollution. In the absence of coordinated government efforts, including stricter enforcement, this figure is likely to rise in the coming year due to the sheer increase in number of vehicles and industries.

REFERENCES

Arora, B.B. (2002) Environmental Pollution, Book : ABC of Biology pp. 839-845.

A.K. Sabharwal 2002. Modern publishers, Jallandhar.

Keshri, D.K. (2006) Carbon monoxide pollution in India & its effect on the health of human being. Book : Biodiversity and Environment pp. 213-216 APH Publishing Corporation, New Delhi.

Pandey, D.K. (2001) Effects of Air Pollutants on health. Emp. News Vol XXVI No. 33, 2001.

Verma, P.S. and Agrawal, V.K. (1998) Pollution Book : Environmental Biology pp. 498-499 S. Chand & Co. Ltd. New Delhi, 1998.

The Effect of Different Concentration of Folidal on the Glycogen in Liver of Snakehead Fish *Channa punctatus* (Bloch)

*P. Dubey**

SUMMARY

Pesticides are being used world wide on large scale in these days. In India 31930 tons are used during 2001 to 2003 according to S. Marain 2003. These concentrations may sometimes reach to lethal levels causing fish mortality, which has harmful effect on human being when consumed by them. The effect of Folidal at three different sub-lethal doses, 100 ppm, 200 ppm and 300 ppm was studied on total glycogen in full-grown specimen of *Channa punctatus* (Bloch). The glycogen contents were estimated. The total glycogen is significantly decreasing compared to control group. The change may be due to glycogen breakdown to glucose under the pesticide toxicity.

Key words : Pesticide, Folidol, Liver Glycogen, *Channa punctatus*.

INTRODUCTION

Pesticide of different types is used to control pests in crops, vegetables and plants. They have been used since about 1850 but before 1940s they were used in relatively small amounts and caused no harmful effects. Recently a great deal of attention has been paid to evaluate the hazardous effect of organ

* Dept. of Zoology, T.N.B. College, T.M.Bhagalpur University, Bhagalpur-812 007 Bihar, India.

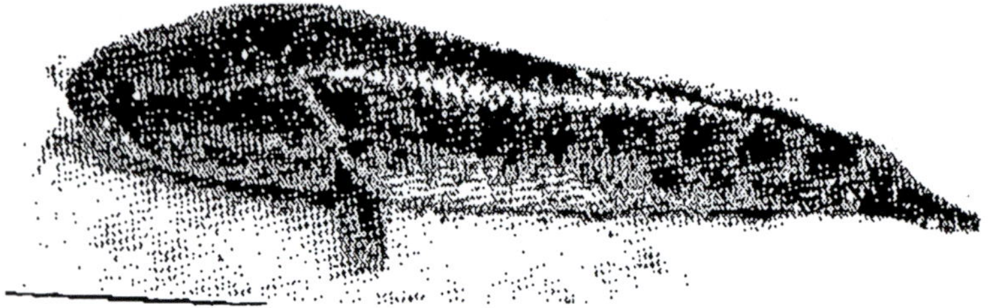

Fig. 14.1: Channa punctatus

phosphorus compound on physiology of many non-largest organism, particularly, fish. The pesticides pollute the aquatic ecosystem were carried by rain water from have they pass into the food chain, ultimately produced toxicity to fishes, birds, wild life and man. The present study evaluates the toxicity of different concentrations of a common pesticide, folidol on glycogen in Snakehead Fish *Channa punctatus* (Bloch).

MATERIAL AND METHODS

The fish *Channa punctatus* is easily available in the locality in ditches, ponds, swamp and paddy fields. Young specimen of the test fished was selected for the experiments Fig. 14.1. They were acclimatized at the room temperature in glass aquaria (75 cm x 37.8 cm x 37.8 cm) containing 25 liters non-chlorinated tap water to assess the effect of Folidol on the fish. *Channa punctatus* (Bloch). The LC_{50} value has been calculated by log dose probit regression line method (Finney, 1971). The glycogen content was estimated by the method of Rex Montgommery (1957).

RESULT AND DISCUSSION

Total glycogen of the liver in control group ranged from 0.024 to 0.032 with an average of 0.001 mg/dl. The liver total glycogen after intoxication of folidal at 100 ppm for 24 hours ranges from 0.004 to 0.036 with an average of 0.017 mg/dl after 48 hours ranges from 0.0016 to 0.024 with an average of 0.016 mg/dl, after 72 hours ranged from 0.004 to 0.028 with an average of 0.016 mg/dl, while after 96 hours ranges from 0.0016 to 0.033 with an average 0.012 mg/dl, at 200 ppm for 24 hours ranges from 0.02 to 0.028 with an average of 0.024 mg/dl, while after 48 hours range from 0.02 to 0.016 with an average of 0.016

mg/dl, while after 72 hours ranges from 0.0016 to 0.032 with an average 0.008 mg/dl. At 300 ppm for 24 hours ranges from 0.032 to 0.216 with an average of 0.025 mg/dl, while 48 hours ranges from 0.002 to 0.24 with an average 0.0092 mg/dl, while 72 hours ranges from 0.0008 to 0.012 with an averages 0.0049 mg/dl, while after 96 hours ranges 0.0004 to 0.002 with an average 0.0017 mg/dl.

The decrease in liver total glycogen of exposure time of folidal, compared to control in no significant (D>0.05) at the 24, 48 and 72 hours but very highly significant (P<0.01) at 96 hours at the 100 ppm of sub-lethal concentration, while 200 ppm of sub-lethal concentration non significant (P>0.05) at 24 hours and highly significant at 48 hours and very highly significant (P<0.001) at 72 and 96 hours, while at 300 ppm of sub-lethal concentration at 24 hours non-significant (P>0.05) at 48 hours significant (P<0.05) 72 and 96 hours is very highly significant (P<0.01) was observed.

The liver glycogen shows decreasing trend on exposure of folidal at different time intervals (24, 48, 72 and 96 hrs.) at 100 ppm sub-lethal concentration. The similar trend was observed at 200 ppm and 30 ppm sub-lethal concentration. The fall in glycogen content in the tissue indicates its rapid utilization by the respective tissue as a consequence of pesticides toxic stress due to glycogen break down to glucose.

At last this type of finding gain support with the finding of Cope et al. (1970). Similar type of decrease in tissue glycogen content of the liver was reported in Tilapia treated with malathion by Kabeer Ahmad et al. (1990), since glycogen depletion is were prevented under hypoxia condition, Dezwaan and Zandee (1995) and Devi (2005).

REFERENCES

Cope. O.B. et al., 1970. Some chronic effect of 2, 4-D on blue gills. Laponis macrochurus. Trans. Am. Fish. Soc. 99.1.

Dubey, P., (1992). Ph.D.Thesis, T.M. Bhagalpur University.

Devi, A.P. et al. 1982. The effect of endosulfar and its isomer on tissue protein, glycogen and lipids in the fish Channa punctatus. Pestic, Biochem. Physio,-17(3): 282-296.

Dezwaan, A and Zandee, D.I. 1972. The utilization of glycogen and accumulation of some intermediate diving analrobiosis in Mytius edulish. comp. Biochem. Physiol, B. 43, 47-54.

Kabeer, A.S. and K.V. Ramana Rao, 1980. Toxicity of Malathion to the fresh water fish, Tilapia mossambica. Bull. Envi.contam. Toxico 1:24:870-874.

Mantgommery, R. 1957. The determination of glycogen, Arch. Biach. Biophy., 67:378-386.

Protective Role of Ascorbic Acid on the Lead and Cadmium induced Alteration in total R.B.C. of the Fresh Water Fish *Channa orientalis*

V.R. Borane[1] and S.P. Zambare[2]

SUMMARY

Fresh water fishes, *Channa orientalis* were exposed to chronic dose of $PbCl_2$ and $CdCl_2$ with and without ascorbic acid. Total erythrocyte counts were recorded. Remarkable decrease in total erythrocyte count was observed in Pb and Cd exposed fishes and the impact was more in Pb as compare to Cd exposed fishes. Fishes were exposed to heavy metals with L-ascorbic acid showed less present variation on the total erythrocyte count. Pre-exposed Fishes to heavy metals showed fast recovery with ascorbic acid as compared to these cured naturally.

Key words : Lead, Cadmium, Ascorbic acid, Erythrocyte, *Channa orientalis*

INTRODUCTION

Trace amounts of metals occur naturally in water, however wastewater from mining, chemical industries and from agriculture etc. are the main source of pollution. Additionally rain contaminated by burning fossil fuels is another source of pollution (Pitter, 1999). In fish, the route of heavy metals entry is either through gills or through mouth. The blood carries

[1]Jijamata Arts, Science, and Commerce College, Nandurbar 425412, India.

[2]Dept. of Zoology, Dr. B. A. Marathwada University, Aurangabad 431004, India.

these heavy metals to different organ or system hence as blood components are directly affected can be well understood by analyzing either blood or serum. Hematological studies have long been considered as a valuable diagnostic tool, in clinical biochemistry, population, physiology and in genetics. For many years it has been used as a diagnostic tool to investigate diseases and physiological or metabolic alteration (Bansal et al., 1979). Physico-morphological changes in blood indicate the changes in the quality of the environment and therefore blood parameters are important in diagnosing the functional status of the animal exposed to toxicants.

Lead ranks second in the list prioritized hazardous substances issued by U.S. ATSDR in 1999. Exposure to lead has been known to effect human health as it leads to mild hyper chronic and microcytic anemia. Lead a biologically non-essential metal is relatively abundant in nature extensive use in modern times (Todd et al, 1996) The built up and transportation of Pb in water, atmosphere and sediment result in bio-accumulation of the metal in various pockets of food chain. The absorbed Pb enters the blood stream where over 90% is bound to the red cells with a biological half life of 25-28 days (Azar et al., 1975). The biochemical effect of Pb is its interferences with heme synthesis leading to hematological damage (Awad, 1997). The prenatal toxic effects of lead on R.B.C. have recently been studied by scanning E.M. (Dey et al., 1999).

Cadmium is not an essential element for organism but it is highly toxic for organisms that live in the aquatic environment. A source of cadmium contamination from release of industrial waste water most significant cases of long term contamination are recorded when cadmium is deposited in bottom sedimentation (Leontovicova, 2003). The heavy metals produce cumulative deleterious effects on animals, which affects on the R.B.C. count and Hb content. However studies on the vitamin C against Pb and Cd toxicity in animal system is limited and in conclusive (WHO, 1980). Antioxidant can play significant role in the treatment of metal induced oxidative stress. Some antioxidants behave as efficient chelators (Gurer and Eracel, 2000). The compound like Vitamin (Vit. C, Vit. E), essential metals (Zinc and Selenium), organic compound (lipoic acid) posses strong antioxidant property. Ascorbic acid has reversed dysfunction of cells lining blood vessels. The normalization of functioning of these cells may be link to prevention heart diseases (Chambers, 1999). It has been found that blood glutathione and AA supplementation were the most aminolevulinic acid dehydratase (Satsangi and Dua, 2000). No doubt, the protective relationship between ascorbic acid on lead and cadmium has been investigated by several workers in certain vital organs of mammalian systems. However, scant attention has been paid on blood parameters. In animals ascorbic acid content in the tissues increases in stress condition during metal toxicosis. It indicates positive role of ascorbic acid in detoxification. Mahajan and Zambare (2001) found recovery in the metabolism altered by heavy metal stress in bivalves. Hence this study was undertaken to assess the role of ascorbic acid on some R.B.C. of an experimental fish, *Channa orietilalis* after chronic exposure to lead and cadmium.

MATERIALS AND METHODS

Medium sized fresh water fishes *channal orientalis*. It (length 6-8 c.m. and weight 25-35 g.) were collected from Shivan River from Nandurbar district and acclimatized in the laboratory condition in well aerated dechlorinated for 8-10 days. Fishes were washed with 0-1% $KMNO_4$ solution to avoid dermal infection The physico-chemical parameters of the water used for the maintenance of the fishes were analyzed as per the methods given in APHA and AWWA (1985). The fishes were divided into three groups A, B and C. Group A fishes were maintained as a control. The Group B fishes were exposed to LC_{50} 10 dose of Pb^- (2.867 ppm) as $PbCl_2$ Cd^- (1.248 ppm) as $CdCl_2$ for 45 days, while group C fishes were exposed to respective chronic concentration of heavy metal with 50mg/l. of ascorbic acid for 45 days. Fishes from B groups were divided into two groups after 45 days exposure to heavy metal into D & E groups. Fishes of D groups were allowed to cure naturally while those of E groups were exposed to ascorbic acid. Total RBC count were recorded from A, B and C group fishes after 15, 30 and 45 days of exposure and from D and E groups after 50th and 55th days of recovery.

Blood was obtained by cutting the caudal peduncle dissection method (Roberts 1978; Reichenbach-Klinke 1982), using heparin as anticoagulant. First few drop were discarded and only the first 2 ml. of blood was taken since the entry of lymph into the blood is reported (Schercher 1954) to affect haematocrit value. The blood was transferred to empty injection glass vials containing anticoagulant in requisite-quantity. RBC was counted by Neubauer haemocytometer using Hayem's and Tuerk's solution respectively as dilution fluid. RBC counts (TEC) were assessed.

RESULTS AND DISCUSSION

The data obtained regarding the physico-chemical properties is given in Table 15.1 while the total RBC counts after exposure to lead and cadmium with and without ascorbic acid and during recovery are given in the Tables 15.1 to 15.2. After chronic exposures to $PbCl_2$ total RBC counts remarkable decreases as compare to $CdCl_2$ were found. In the presence of ascorbic acid (50 mg/l) the total RBC counts decreases less as compared to those of heavy metal intoxificated fishes. The fishes pre-exposed to heavy metals salts showed fast recovery in the RBC counts in the presence of ascorbic acid, than those allowed to cure naturally. The treatment with lead and cadmium decrease RBC as compared to control. Whereas, supplementation of AA to lead intoxicated rats exhibits reducing influence of lead toxicity. Significant decrease in RBC as a result of lead intoxicated might be due to hemolytic anemia under the influence of toxic metal. (Singh el al., 2005). The administration of lindane to male rats at different dose levels for 30 days produces significant changes in haemopoietic system and in seaim biochemical parameters (Joshi et al., 2005). The decrease in RBC counts may be due to either inhibition of RBC production or destruction of RBC due to poisoning leads to fall in Hb concentration (Reeves el al., 1981; Siddiqui et al., 1987).

Jezierska and Witeska (2001) reported that mucus secreted in large amounts by metal affected gills reduced membrane transport of oxygen in the respiratory epithelium and that metal accumulated in this epithelium disturbed the oxygen diffusion. Drastichova (2004) reported that cadmium in the

Table 15.1 : Physico-chemical parameters of water used for experimentation

Temperature	25-1 ±3-2ᵘ
pH	7.60±0.3
Conductivity	140±15.7 μ mho-cm
Free CO_2	3-34 ±1- 3ml^{-1}
Dissolved O_2	6-3 ± 1-1m^{-1}
Total Hardness	204 ± 12.0mg^{-1}.
Total Alkalinity	585-6±32.8mg^{-1}
Magnesium	31-67±2.9mg^{-1}
Calcium	30-46 ±3.06 mg^{-1}
Chloride	107-92 ± 16.34 mg^{-1}

form of $CdCl_2$ was classified among toxic substance for fish. Changes in the R.B.C. profile suggest compensation of oxygen deficit in the body due to gill damage. Nature of the changes shows a release of erythrocytes from the blood. A decreased HC+, Hb and R.B.C. in *Aphanius dispar* after Hg exposure has been attributed to the permeability of R.B.C. to mercurial compounds and the balance of the number of binding sites between intracellular extracellular components of blood. (Shah and Altindag, 2004). Vitamin C not only confer protection against lead toxicity, but it can also perform therapeutic role against such toxicity in general (Bhattacharjee et al., 2003). The decrease in the rate of R.B.C. count and hemoglobin contents indicates the impact of heavy metals on the erythropoietic tissue. It is well known that the cellular respiration in vertebrates depend on the availability of iron associated with Hb and hence depends on R.B.C. in its intact form it acts as an efficient oxygen carrier. Ascorbic acid also reported to reduce the levels of Pb in blood, liver and kidney during Pb exposure in rats The ascorbic acid acted as detoxifying agent by forming poorly ionized, but soluble compound with lead (Pillemer, 1940). Data on toxicity of heavy metals and their effects on aquatic organisms are the basic ones for determination of eco-toxicological risks of heavy metals for the aquatic ecosystem. The present paper contributes to assessment of toxicity and effect of lead, cadmium,

Table 15.2: R.B.C. counts in *Channa orientalis* after chronic exposure to $PbCl_2$ and $CdCl_2$ without and with ascorbic acid (millions/mm^3).

Group	Treatment	15d	30d	45d	50d	55
A	Control	3.3±0.125	3.2±0.17	3.00±0.294	--	--
B	Pb^{--} (2.867ppm)	2.3±0.057** (-30.30)	2.00±0.21* (-37.5)	1.65±0.092* (-45)	--	--
C	Pb^{--} (2.867ppm)+ AA	2.63 ±0.037** (-20.30)	2.5 ±0.081* (-21.87)	2.13 ±0.047* (-29)	--	--
D	Recovery in Normal water	--	--	--	1.92±0.026$^{\Delta}$ [+16.36]	2.10±0.014$^{\Delta\Delta}$ [+27.27]
E	Recovery in AA	--	--	--	2.19±0.143$^{\Delta}$ [+37.72]	2.3±0.163$^{\Delta}$ [+39.39]
B	Cd^{--} (1.248ppm)	2.6±0.139* (-21.21)	2.3±0.129* (-28.12)	1.91±0.081* (-36.33)	--	--
C	Cd^{--} (1.248ppm)+ AA	2.71 ±0.130* (-17.87)	2.61 ±0.082* (-18.43)	2.4 ±0.065* (-20)	--	--
D	Recovery in Normal water	--	--	--	2.29±0.018$^{\Delta\Delta}$ [+19.89]	2.50±0.015$^{\Delta\Delta}$ [+30.89]
E	Recovery in AA	--	--	--	2.49±0.129$^{\Delta}$ [+30.36]	2.70±0.02$^{\Delta\Delta\Delta}$ [+41.36]

AA = Ascorbic acid (50 mg/1., ± indicates S.D. of three observations.
Values in ( ) indicates percent change over respective control.
Values in [ ] indicates percent change-over 45 days of respective B.
* indicates significance with the respective control.
Δ indicates significance with 45 days of respective B. $p<0.05$ = *&Δ $p<0.01$ = **&ΔΔ, $p<0.001$ =***& and ΔΔΔ = Not significant.

and ascorbic acid on toxicosis. The aim was to assess the effect of lead and cadmium on the oxidative blood capacity and the non specific immune response by means of examination of R.B.C. count. Changes indicating to adaptation in the red blood cell system including the decrease in R.B.C. were probably a response to impairment of gas exchange in lead and cadmium affected gills and recovery in ascorbic acid and naturally cured.

REFERENCES

APHA, AWWA and WPCF (1985): Standard methods for the examination of water and waste water APHA (17th) INC, New York.

ATSDR (1999): Agency for Toxic substance and Disease Registry.

Awad M William, Jr. (1997): Textbook of biochemistry with clinical correlations John Wiley and Sons, INC, New York.

Azar A, Snee RD and Habibi K (1975): Lead toxicity in mammalian blood In : TB Griffin, J.H. Knelson (ed.) Lead Academic, New York, p. 254.

Bansal S.K., Verma S.R., Gupta A.K. and Dalela R.C. (1979): Physiological dysfunction of the haemopoietic System in a fresh water teleost *Labeo rohita* following chronic chlordane exposure. Bull. Environ. Contam. Toxicol. 22: 674-680.

Bhattacharjee C.R., Deys S. and Goswami P. (2003): Protective role of ascorbic acid against Pb toxicity in Blood of albino mice as releaved by metal up take lipid profiles and ultra structural features of Erythrocytes Bull Environ. Contam. Toxicol. 70: 1189-1196.

Chambers J.C., Greger M.C. and Jean Marie (1999): Demonstration of rapid onset vascular endocrine dysfunction after hyperhomocy teinemia. An effect reversible Vit. C therapy. Circulation, 99:1156-60.

Dey S, Arjun J, Das M (1999): Erythrocyte membrane dynamics in albino mice offspring born to female with inducted lead toxicity during pregnancy. A scanning electron microscopic study. Biomed Lett 59: 55-66

Drastichova J, Svobodova Z., Lusko and V., Machova J., (2004): Effect of cadmium on haematological indices of common carp *Cyprhnis carpio* L. Bull. Environ. Contam. Toxicol. 72:725-732.

Gurer H. and Eracel N. (2000): Can antioxidant be beneficial in the treatment of lead poisoning? Free Rad. and Med. 29 (10): 927-945.

Jezierska B and Witeska M (2001) Metal toxicity to fish Wydawnictwo, A.P. sidlece.

Joshi SC, Goyal Rekha, Choudhary Nisha, Jain Sunita (2005) Effect of lindane on haematological and serum parameters of male albino rats. National J. life Sciences, 2 (Supp.): 227-230.

Leontovicov'a D (2003): Complex monitoring in selected profiles of state networks of water quality follow up in CHMU Periodlcum fakulty ekologie a enviromentalistiky Technickej university Yozvolene vol. 10 Suppl.

Mahajan A.Y and Zambare S.P. (2001): Ascorbate effect on $CuSO_4$ and $HgCl_2$ induced alteration of protein levels in fresh water bivalve *Corbicula striatella* Asian J. Microbiol Biotech and Environ Sci. Vol. 3, No (1-2): 95-100.

Pillemer L.J, Seifter A.O., Ecker, E.E. (1940): Vitamin C in chronic lead poisoning. An Experimental study. American J. Med. Sci. 200: 322-327.

Pitter P. (1999) Hydrochemie (Hydrochemistry) Vydavatelstvi VSCHT, Praha.

Reichenbach-Klinke H.H. (1982): Enfermedades de los peces. Ed Acribia, Zaragoza. Espana, p. 507

Reeves J.D., Driggers DA, and Vincent A.K. (1981) Child Leukemia and A plastic anaemia after malathion exposure. The Lancent 300.

Roberts R.J. (1978): Fish Pathology. Bailliere Tindall. New York, USA, p377.

Satsangi Kiran and Dua K.K. (2000) Preventive effects of few dietary nutrients against acute aluminium toxicity in mice. Trace Elements and Electriolytes, 17(3): 134-137.

Schermer S. (1954) Die Blutmorphologie der laboratorium-stiere Earth, Leipzig Experta Medica Foundation, FA Davis Co. Philadelphia.

Shah S.L., Altindag A (2004) Haematological parameters of Tech. (*Tinca tinca* L.) after acute and chronic exposure to lethal and sublethal mercury treatments Bull. Environ Contam. Toxicol. 73: 911-918.

Siddiqui M.K.J, Rahaman M.F., Anjuman F. and Qadri S. (1987): Administration of endosulfan on some haematological parameters and serum enzymes in rats. Pesticides, pp. 25-27.

Singh Sushitima, Chaturvedi S. and Gaur K.K., Singh A. (2005): Role of ascorbic acid on blood in lead intoxicated albino rats *Rattus norvegicus*. Nat. J. Life Science, 2:343-344.

Todd A.C., Wetmur J.G., Moline J.M., Godbold J.H., Lewin S.M., Landrigen P.1 (1996) Unvavelling the Chronic toxicity of lead. An essential priority for environmental health environ. Health Perspect. 104(suppli): 141-146.

WHO (1980) Laboratory manual for the examination of human semen semal cervical mucus interaction Geneva.

Influence of Fenvalerate on Ascorbic Acid Content in the Tissues of Fresh Water Bivalve *Parreysia cylindrica*

R.D. Patil and S.R. Magare***

SUMMARY

Synthetic pyrethroids cause serious pollution problem in aquatic organism environment. The ascorbic acid content of mantle foot gill digestive gland and whole body of *Parreysia cylindrica* was studied after acute and chronic exposure to fenvalerate. The bivalves were exposed to 8.964 ppm and 2.712 ppm to fenvalerate as acute and chronic treatment. There was a significant decrease in the ascorbic acid contents in mantle foot, gill, digestive gland and whole body tissues.

INTRODUCTION

Fenvalerate is a synthetic pyrethroid known to be lipophilic and of potential hazard to aquatic ecosystem (Day and Kaushik 1987; Elliot 1977). Pesticides form the major part of toxicants polluting the aquatic environment. The transportation of the pesticides into aquatic environment via different agencies posses the problem of contamination and mostly the non-target organism are often affected (Bradbury et al., 1984, Sangeet Kumar et al. 1990; Sampath et al., 1992)

Ascorbic acid is an essential factor responsible for collagen biosynthesis. It is essential for the normal development and in

* Department of Zoology, Arts, Commerce and Science College, Navapur, Dist. Nandurbar - 425418 , (M.S.)

** Department of Zoology, A.A. Mandal's C.H.C. Arts S.G.P. Commerce and B.B.J.P. Science College, Taloda, Dist Nandurbar (M.S.) 425418.

most species it is produced endogenously thus resembling a hormone. Glucose and other hexoses are convertible to glucose serve as the starting material for biosynthesis of ascorbic acid. Halver (1972) states that the ascorbic acid plays a major role in tissue synthesis and growth process obviously mediates rapid tissue repair in trauma or disease condition. A major function of ascorbic acid is the formation of tissue collagen, also it take part in the maturation of red blood corpuscles. (Talwar, 1980).

Very little information exists regarding the influence of pyrethroids on the ascorbic acid content in the various tissues of the bivalve. Changes involved in the ascorbic acid content in the tissue of freshwater bivalve *Parreysia cylindrica* exposed to pyrethroid pesticide fenvalerate was studied.

MATERIALS AND METHODS

Healthy and nearly equal size freshwater *bivalves, Parreysia cylindrica* were collected from Girna dam, 44 km away from Chalisgaon, Dist. Jalgaon, Maharashtra. They were maintained in the troughs and acclimatized for 3-4 days in the laboratory. The animals were grouped into control and experimental sets.

In acute treatment experimental bivalves were exposed to 8.964 ppm (LC 50/2 ppm of 96 hours) of fenvalerate. At the end of 24, 48, 72 & 96 hours of treatment control and treated animals were dissected and their mantle, foot, gill and digestive glands were separated and whole body mass was taken. All tissues were dried at 70°C to 80°C in an oven till constant weights were obtained.

For chronic treatment the experimental bivalves were exposed to 2.712 ppm (LC 50/10 ppm of 96 hours) of fenvalerate upto 21 days. After 7th, 14th and 21st days of treatment control and treated bivalves were dissected and their mantle, foot, gill and digestive glands were separated and whole body mass of remaining animals was taken. All tissues were dried at 70°C to 80°C in an oven till constant weights were obtained.

The dried powders of different tissues of control and experimental animals were used for estimation of ascorbic acid content by hydrazine reagent method (Roe 1967). The amount of ascorbic acid content was expressed in terms of mg of ascorbic acid/100 mg of dry weight of tissues. Each observation was confirmed by taking at least three replicates. The difference in control and experimental animal group was tested for significance by using student 't' test (Bailey, 1965) and the percentage of decrease or increase over control was calculated for each value.

RESULTS AND DISCUSSION

In *Parreysia cylindrica,* the ascorbic acid content in the mantle, foot, gill, digestive gland and whole body was decreased after acute and chronic treatment of fenvalerate (Tables 16.1 and 16.2)

Gould (1963) reported acclimation of ascorbic acid at the site of wound healing. The increased demand of energy being provided by utilization of ascorbic acid in response to pesticide stress (Gorbunova, 1966; Chinoy, 1972a, b). Mc Cann and Jasper (1972) reported hemorrhage and vertebral injuries in fishes exposed to high level of various pesticides. Stress caused alteration in the normal physiology of animal leading to enhanced utilization and mobilization of ascorbic acid (Chinoy and Kamalakumari, 1976), ascorbic acid is recognized as antistress factor (Kutsky, 1973). In acute physiological conditions the ascorbic acid content has been shown to decrease in brain and gut and shift to muscle and other body parts in *Channa gachua* due to increased demand of energy on fatigue (Ali et al.; 1983). Decrease in ascorbic acid content in whole body, mantle, foot, gill and digestive gland of *Corbicula striatella* after exposure to carbaryl was reported by Jadhav et al., (1996) and stated possible utilization of ascorbic acid to overcome the stressful condition.

REFERENCES

Ali S.M., and Ilyas R. and Bhusari N.B. (1983): Effect of lethal and sublebhal concentration of dimecron on ascorbic acid of fish, *Channa gachua (Ham.) Marathwada University Journal of Science,* XII (15): 69-74.

Bailey N.T.J. (1965). Statistical methods in Biology. ELBS. English University Press, London.

Table 16.1: Fenvalerate induced changes in the ascorbic acid content in different tissue of *Parreysia cylindrica* after acute exposure

Tissue	*Control*	*Fenvalerate Treated* 24 Hrs.	48 Hrs.	72 Hrs.	96 Hrs.
Mantle	1.28374±0.1321	0.9446 + 0.0164±	0.8219±	0.7062±	0.6522
	-26.41*	0.0060	0.0276	0.0465	
		-3597**	-44.98 **	49.19	
Foot	0.6694 + 0.0556	0.5833±0.1135	0.5415±	0.5290±	0.4161±
		- 12.86 NS	0.0371	0.0456	0.0599
–19.10*	–20.97*	–37.83**			
Gill	1.6918± 0.1040	1.5217±0.1215	1.4005±	1.2941±	0.9568±
	- 10.05 NS	0.0173	0.0583	0.2310	
		- 17.21 *	-23.50**	. 4344**	
Digestive	1.835±0.0208	1.2326±0.0903	1.0881±	0.8411±	0.6503±0.02
–32 82 ***	0.1248	0.0420	–6456***		
		-40.70***	-54.16***		
Whole Body	1.5178±0.0465	1.2149±0.0382	0.9100±	0.7245±	0.5982±
	- 17 97 ***	0.0173	0.0152	0.0081	
		-40.04***	-52.26***	-60.58***	

Values expressed as mg / 100 mg dry wt. of tissue
+ or - , % variation over control.
± indicate S.D. of three observations.
Significance : * P < 0.05 ; ** P <0.01 ;*** P<0.001; NS = Not Significant.

Table16.2: Fenvalerate induced changes in the ascorbic acid content in different tissue of *Parreysia cylindrica* after acute exposure

	Control			*Fenvalerate Treated*		
Tissue	*7 Days*	*14 Days*	*21 Days*	*7 Days*	*14 Days*	*21 Days*
Mantle	1.2554± 0.0544	1.2314± 0.0282	1.2301± 0.0439	0.9409± 0.0141 -25.05***	0.8743± 0.0081 30.35***	0.8512± 0.01 -32.19***
Foot	0.6692± 0.0355	0.6544± 0.0173	0.6351± 0.0081	0.6083± 0.0057 -9.10*	0.4478± 0.0115 -33.08***	0.3815± 0.0129 -42.99 ***
Gill	1.5534± 0.0695	1.5929± 0.0896	1.5300± 0.0244	0.9801± 0.0115 –36.90 ***	0.7221± 0.0086 51.58***	0.4879± 0.01 –68.59 ***
Digestive Gland	1.6197± 0.0036	1.5201± 0.0768	1.4865± 0.0529	0.8522± 0.0138 - 47 38 ***	0.5683± 0.0251 -64.91 ***	0.3006± 0.0157 -81 44 ***
Whole Body	1.2548± 0.0734	1.2215± 0.1122	1.2034± 0.1244	1.0138± 0.0360 –19.20**	0.8388 + 0.0206 –33.15**	0.5107 + 0.0104 –59.30***

Values expressed as mg / 100 mg dry wt. of tissue
+ or - , % variation over control.
± indicate S.D. of three observations.
Significance : * P < 0.05 ; ** P <0.01 ;*** P<0.001; NS = Not Significant.

Bradbury, S.P. Jole R. Coats James M. Mekim (1984). Differential toxicity and uptake of two fenvalerate formulation in fat head minnows *(Pimples Promlas)* Environ. Toxicol. Chem - 4, 533-541.

Chinoy, N.J. (1972, a) : Ascorbic acid levels in avian tissues and its metabolic significance. Acta. Zool. (Stockh), 53 (1) : 121-125.

Chinoy, N.J. (1972, b) : Ascorbic acid level in mammalian tissues and its metabolic significance. Comp. Biochem. Physio. 42 A: 945-952.

Chinoy, N.J. and Kamalakumari, D. (1976) M.Sc. Dissertation, Zoology dept. Gujarat University, Ahmedabad.

Day K and Kaushik N.K. (1987) : The adsorption of fenvalerate to laboratory glassware and the alga *Chlamydomonds reinhardii* and its effect on uptake of the particide by Daphnia *Galeata mendolae.* Aquatic Toxicol. 10:131-142.

Elliot M (1977) : In synthetic pyrethroid Ed. : Elliot M American Chemical Society Washington D.C. pp. 1-28.

Garbunova, T.V. (1966) In : Directed Biosynthesis, Nauka, Moscow.

Gould B.S. (1963) Internal Review of Cytology (5) Academic Press New York.

Halver, J.E. (1972) : The role of ascorbic acid in fish disease and tissue repair . Bull. Jap. Soc. Sci. Fisheries 35 (i): 79 - 92.

Jadhav, Sunita Y.B. Sontakke and V.S. Lomte (1996) : Effect of Carbaryl on Ascorbic acid content in the selected tissues of *Corbicula striatella.* J. Ecotoxicol. Environ. Monit. 6(2) : 109-112.

Kutsky R.J. (1973) Handbook of vitamins and hormones van nostrand rainhold, New York.

McCann and R.C. Jasper (1972) vertebral damage to blue gills exposed to acute tork levels and pesticides. Trans. Amer. Fist Soc. 101-307.

Roe, J. (1967) In : Methods of biochemical analysis Vol. 5 Ed. Blick Inter Science, New York pp:44.

Sampath, K, Elango, P and Thanalaxmi, S. (1992): Effects of carbaryl (sevin) on the carbohydrate metabolism of the common frog *Rana tigirina,* Environ. Ecol. 10(2) ; 278-281.

Sangeet Kumar Varma, Varma, K. Pathak, S and Anil Kumar (1990) : Effects of falling environmental temperature on the lipid content of the fat body. *Rana tigrina* Environ. Ecol. 8(1), 509, 511.

Talwar G.P. (1980) : Text book of Biochemistry and human biology, prentice - Hall of India, New Delhi 1042 pp.

Toxic Effects of Aqueous Extract of Leaves of *Cestrum noctunum* (Linn.) on Respiration of the Freshwater Gastropod *Lymnaea luteola*

R.D. PATIL[1] AND S.R. MAGARE[2]

SUMMARY

The present chapter deals with the effect of aqueous extract of leaves of *Cestrum nocturnum* (Linn.) on respiration of the freshwater gastropod *Lymnaea Luteola*. Individuals were exposed to 29.063 mg ($LC_{50/2}$) aqueous extract of *Cestrum nocturnum* as acute treatment and 5.812 gm ($LC_{50/10}$) as chronic treatment.

The rate of oxygen consumption was decreased in the experimental animal compared to individuals in control.

Key words: Oxygen consumption, *Cestrum nocturnum*, *Lymnaea luteola*.

INTRODUCTION

Plants are an integral part of nature. A number and plants have been reported as stupefying and piscicidal agents (Jhingram, 1983 ; Mahajan, 1994). *Cestrum nocturnum* is an ornamental plant from Solanaceae family. It is found to be highly toxic to fishes though its toxicity is reported in terrestrial animal i.e. goat and buffaloes (Durand et al.; 1999). Histomorphological changes in gill of fish *Nemachaeilias evezardi*

[1]Department of zoology, Arts, Commerce and Science College, Navapur, Dist.Nandurbar - 425418, (M.S.) India.

[2]Department of zoolpgy, A.A. Mandal's C.H.C. Arts S.G.P. Commerce and B.B.J.P. Science College, Taloda, Dist. Nandurbar (M.S.) 425418.

after exposure to aqueous extract of leaves of *Cestrum nocturnum* has been reported by Patole and Mahajan, 2004)

Oxygen is essential to provide energy for various life process. A good deal of information is available on the oxygen consumption of crustaceans, gastropods and fishes but relatively less work has been done on the oxygen consumption of freshwater molluscs; (Hiscock, 1953, Rotauwe, 1958 and Ramamurthi, 1965). The relationship between respiratory activity of the animal and pollution has been studied in bivalves and gastropods (Lomte and Jadhav, 1982; Zambre et al.; 1996)

The information regarding the effect of *Cestrum nocturnum* on respiratory metabolism in *Lymnaea luteola* is lacking. Hence the present study has been undertaken to study the alteration in oxygen consumption after pesticidal exposure in *Lymnaea Luteola.*

MATERIALS AND METHODS

Medium sized live and healthy freshwater gastropod *Lymnaea luteola* were collected from local freshwater source and they were acclimatized to laboratory condition for 3 days. Plant material was collected from Navapur area, District Nandurbar. Leaves were dried at room temperature and powdered using grinder. Various aqueous extracts were prepared by treating the material with water. Toxicity of the leaves of test plant was evaluated by static bioassay method and LC_{50} values were evaluated as described by Mahajan et al. (1989).

Healthy and active snails were taken 40 each in number in three different troughs. One trough was maintained as control, animals in second trough was exposed to 29.063 mg ($LC_{50/2}$) aqueous extract of leaves of *Cestrum nocturnum* as acute treatment and the animals of third group were exposed to 5.812 gm ($LC_{50/10}$) aqueous extract of leaves of *Cestrum nocturnum* as chronic treatment.

The rate of oxygen consumption of control animals exposed to acute treatment was studied after every 24 hours upto 96 hours and that of animals exposed to chronic treatment was studied every seven days with control for 21 days. Oxygen consumption was estimated by Winkler's method (Welsh and Smith, 1960). The "t" test was carried out and percent change in oxygen consumption was noted. No special food was provided during experiment.

RESULTS AND DISCUSSION

The rate of oxygen consumption in the *Lymnaea luteola* was decreased after acute and chronic treatment of aqueous extract of leaves of *Cestrum nocturnum* (Tables 17.1 and 17.2).

The alcoholic extract of aerial part of *Cestrum nocturnum* was found to have CNS represent action in mice (Chatterjee and Roy, 1964). The leaves of *Cestrum noturnum* contain Tigogenine, Yuccagenin, saponin, alkaloids, Nomicotine, Continine, Myosmine and essential oil (Karawya et al., 1971). The primary action of pesticide badly damages the gill surface and reduce the oxygen uptake capacity of the respiratory organs (Roberts, 1978). Saponin of *Cestrum nocturnum* have been reported to have biological activity like cardiotoxic, haemolytic, cardiotonic etc., (Chatterjee and Roy, 1964, Mahajan et al., 1989). The aqueous extract of the plant showed excellent piscicidal activity in fish *Nemachaelius sinyuatus, Punctus Sarana and Lebistus reticulates* (Mahajan et al., 1989). The biocidal effect of aqueous extract of leaves of *Cestrum nocturnum* on the heart beat and cardiotoxicity has been reported by Jawale and Mahajan (1999) in cockroach *Periplaneta americana.*

Depletion in the rate of oxygen consumption in *Nemachaelius evezardi* (Day) after exposure to *Cestrum nocturnum* was reported by Mahajan and Patole (2003). Similar results were obtained in The primary action of pesticide badly damages the gill surface and reduce the oxygen uptake capacity of the respiratory organs (Roberts, 1978). The present study. Decrease in enzyme activity necessary for respiration causes decrease in oxygen consumption. Ruptured gill lamellae, vacuolization and elongated gill lamellae were observed in fish *Nemachaelius everzardi* (Day) after treatment of *Cestrum nocturnum* plant extract (Patole and Mahajan 2004).

REFERENCES

Chatterjee, M.E. and Roy, A.R. (1964) : pharmacological studies with saponin from *Cestrum diurnum,* Bull. Calcutta school of tropical medicine, 12(2):58-62.

Durand, R.J.M. Figured and E. Mendosa (1999): Intoxication in cattle from *Cestrum nocturnum.* Vet. Hum. Toxicology 41(1) : 26-27.

Table 17.1 : Rate of Oxygen Consumption of *Lymnaea luteola* after acute exposure to leaves of *Cestrum nocturnum*.

Sr. No.	*Treatment*	*Average oxygen consumption ±S.D. ml/gm/hr/lit.* 24 Hrs.	48 Hrs.	72 Hrs.	96 Hrs.
		1.9202±	1.911±	1.8717±	1.8356±
1.	Control	0.0251	0.0238	0.0424	0.0301
2.	*Cestrumnocturnnm leaves*	NS	**	***	***
		1.7163±	1.6609±	1.5513±	1.4672±
		0.1832	0.0741	0.0343	0.0270
		10.61	13.08	17.11	20.09

Each Value is mean of three observation + S.D.

Values are significant at * $P < 0.05$, ** $P<0.01$, *** $P<0.001$

NS = Not Significant

Table 17.2 : Rate of Oxygen Consumption of *Lymnaea luteola* after chronic exposure to leaves *of Cestrum nocturnum.*

Sr. No.	*Treatment*	*Average oxygen consumption ±S.D. ml/gm/hr/lit.*		
		7 Days	*14 Days*	*21 Days*
1.	Control	1.8908± 0.03	1.7286± 0.0285 +	1.6434± 0.0199
2.	*Cestrum nocturnum* leaves	*** 1.2547± 0.0783 33.64	*** 1.0744± 0.0542 37.84	*** 0.857± 0.00148 47.86

Each Value is mean of three observation + S.D.

Values are significant at * P < 0.05 , ** P< 0.01 , *** P< 0.001

NS = Not Significant

Hiscock, I.D. (1953): Osmoregulation in Australian freshwater mussel (Lamellibranchiata) II. Respiration and its relation to osmoregulation in *Hydridella australis* (Lam.) Australation, J. Mar. Freshw. Res. 4: 330-342

Jawale S.M. and R.T. Mahajan (1999) : Effect of aqueous extract of *Cestrum nocturnum.* (Linn) on Cockroach : *Periplaneta americana* Proc. Acad. Environ. Biol. 7 (2) : 129-132.

Jhingram, V.G. (1983) : Fish toxicant of plant origin cited in : Fish and Fisheries of India, 2nd Edn Hindustan Publ. New Delhi : 371-372

Karawya, M.S., A.M. Rizk; F.M. Flammouda, FAM Diab. and Z.F. Ahmed (1971) : Phytochemical investigation of certain *Cestrum* species. Planta Med. 20 (4) : 363-367.

Lomte V.S. and M.L. Jadhav (1982) : Effect of toxic compounds on oxygen consumption in the freshwater bivalve *Corbicula regularis* (Prime 1860) Comp. Physiol. Ecol. 7, 31-33.

Mahajan R.T., S.M. Choubey and S.M. Jawale (1989) : Piscicidal activity of some indigenous plant. In perspectives in aquatic biology, papyrus publication, New Delhi : 369 - 375.

Mahajan R.T. (1994) : Review on fish toxicants of plant origin, Ind. J. Environ Toxicol 4(1) : 7-18.

Mahajan R.T. and S.S. Patole (2003) (Day). Advances effect of plant extracts on oxygen consumption rate in fish *Nemachaelius evezardi* (DAY), J. Freshwater Biol. 15 (1-4) : 109-113.

Patole, S.S. and Mahajan R.T. (2004) : Effect of aqueous extract of the leaves of *Cestrum nocturnum* (Linn.) on histomorphological changes in gill of fish *Nemachaelius everardi* (Day) J. Aqua Biol. Vol. 19(2), 135-140.

Ramamurthi R. (1965) : Metabolic responses to osmotic strees in some freshwater poikilotherms. Curr. Sci. 34: 351-354.

Robert R.J. (1978) : Pathophysiology and systematic pathology of teleost In : Fish pathology (Ed. R.J.O.) pp. 55-59, Billiere Tindall London, U.K.

Rotauwe, H.W. (1958) : Undersuchunger zur A atmungas-physiologic and osmoregulation bei mytilus edulis mt. einem kurzenhag uber die bluthonezentration Von dreisena polymorpha in Abhangigks it Von Ellrolytagentdes Aussen Medium. veroft. Inst. Meerestroch Bremerna Nen 5. 143-159.

Welsh, J. and R.I. Smith (1960) : Laboratory exercise in invertebrate physiology. Burgess minneapolles.

Zambre S.P., Sunita Jadhav and S.P. Bhoi (1996) : Effect of cypermethrin on respiration of the fresh water gastropad, *Lymnaea luteola* (Lamark) Proc. Acad. Environ Biol. 5(2), 131-133.

Pollution Studies of Naregaon Nullaha in the Industrial Area of Aurangabad (M.S.) Part-II

D.L. Sonawane, Thete Pournima, K.T. Paithane and Y.K. Khillare

SUMMARY

In the present study, land and aquatic pollution in the industrial area of Aurangabad has been undertaken to make the health awareness in the residents of locality residing around the Naregaon nullaha. Presence or absence of some aquatic fauna has been considered as an indicator in the present study. Seasonal study on variation of floral and faunal abundance in summer, winter and Rainy seasons have also been considered.

Behavioural studies in *Rasbora daniconius* and *Channa gachua* exposed at different concentrations of the industrial effluent are also considered as an important parameter in aquatic pollution studies.

Key words: Behaviour, industrial effluent, *Rasbora daniconius, Channa gachua.*

INTRODUCTION

Pollution means the addition of any foreign material (inorganic, biological or radiological) or any physical change in the natural water which may harm fully affect the living life (human, agricultural or biological) directly or indirectly. It is an undesirable change in the physical, chemical or biological

Department of Zoology, Dr. Babasaheb Ambedkar Marathwada University, Aurangabad-431004 (M.S.).

characteristics of air, land or water that harmfully affect human life.

The increasing industrialization and urbanization has brought a measurable water crisis. Today, most of the rivers of world receive millions of liters of sewage, domestic waste, industrial and agricultural effluent containing substances varying in characteristics from simple nutrient to highly toxic substances.

Keeping this fact in mind present work has been framed out to study the water quality experimentation on water of Nullaha located fifteen kilometers away towards east of the main city of Aurangabad. It flows in north south direction and Aurangabad is located at latitude 19.50 north and longitude 75.23 east. Since the nullaha flows near area, industrial effluents are directly discharged into it. There were casualties reported in Sheeps, bullock, birds by drinking water of Naregaon Nullaha. This incidence has inspired me to concentrate research on such waste water resources. It is hazardous to human beings as well as many animals including fishes.

Direct discharge of industrial effluent in river is one of the main causes of pollution. The pollutant find their way directly or indirectly into food and living tissue (Hamilt and Sundarvadhanam, 1984). Such studies are useful in making comparison with unpolluted water and also provide information on degree of pollution due to industrial effluent. Severe and rapid damage to the aquatic organisms by fastest action of poisoning (Lund, 1971).

Literature review show that in a century back, toxicity tests of various pollutants on aquatic organisms have been described Forbes (1989). The effect of organic pollutants on aquatic biological communities has been reported by Richardon (1978), he further explained that toxic substances or chemicals which are present in the effluent may enter in the food chain through uptake vegetation, fishes, phytoplankton and zooplankton. Several studies showed alarming level of pollution of the rivers due to industrial effluent and sewage, Humtage (1978); Zindoge et al. (1980); Ajmal et al. (1982); Sumusshekar (1985); Sengarand Sharma (1987).

Table 18.1: Seasonal study on variation of floral and faunal abundance (summer season).

Parameters	*Sampling points*									
	1	2	3	4	5	6	7	8	9	10
Algae	+	+	+	+	+	-	+	+	+	-
Aquatic plants	-	-	-	-	-	-	-	-	-	-
Planktons :- 1 . Phytoplankton	-	-	+	+	+	+	+	+	-	-
2. Zooplankton	-	-	+	+	+	-	+	-	+	-
Mosquito eggs and larvae of other animals	+	+	+	+	-	-	-	-	-	-

+ = Present
– = Absent

Table 18.2: Seasonal study on variation of floral and faunal abundance (winter season)

Parameters	*Sampling points*									
	1	*2*	*3*	*4*	*5*	*6*	*7*	*8*	*9*	*10*
Algae	+	+	+	+	+	-	-	-	-	-
Aquatic plants	-	+	-	+	-	-	-	-	-	-
Planktons:- 1. Phytoplankton	-	+	+	+	-	+	-	-	-	-
2. Zooplankton	+	-	+	+	+	+	+	-	-	-
Mosquito eggs and larvae of other animals	+	+	+	+	-	-	-	+	+	-

+ = Present
– = Absent

MATERIAL AND METHODS

After complete survey of the sampling site, the sampling points were fixed, there was a minimum distance of 1 km between two points. The sampling analyses and reporting have been undertaken continuously for three years to come to any conclusion.

The values which are mentioned are mean of seasonal variations, and are taken on the basis of mean of month wise collection of samples.

The fresh water fishes *Rasbora daniconius* and *Channa gachua* were collected from Paithan near Aurangabad and acclimatized to the laboratory conditions for three days in aquaria with aeration. The fish *Rasbora daniconius* and *Channa gachua* were exposed at 100%, 50%, 20%, 10% and 01% concentrations.

Behavioural observation, seasonal study on variation of floral and faunal abundance in Rainy season, winter season and summer season has also been tabulated.

RESULTS

The results of seasonal study on variations of floral and faunal abundance are recorded in tables for Summer Table 18.1, Winter Table 18.2, and Rainy Season Table 18.3.

The behavioural changes of test fish Rasbora daniconius were observed and recorded in Table 18.4. In 0% and 10% concentration of effluent in all sampling points found normal opercular movement. At 20% concentration of sample the fish showed fast opercular movement in the first sampling point. In the remaining sampling points behavioural changes observed normal movement at 50% and 100% concentration of an effluent. Jerky with fast opercular movement and fish was trying to come out of effluent.

Similarly *Channa gachua* showed behavioural changes at different concentration of an effluent and are recorded in Table No. 18.5. The changes observed at 0% and 10% concentration were normal opercular movement in all sampling points. In 20% and 50% dilution show movement of test fish was observed and normal opercular movements. But in 100% concentration

Table 18.3: Seasonal study on variation of floral and faunal abundance (Rainy season).

Parameters	*Sampling points*									
	1	*2*	*3*	*4*	*5*	*6*	*7*	*8*	*9*	*10*
Algae	+	+	+	+	-	-	-	-	-	-
Aquatic plant	+	+	+	+	-	-	-	-	-	-
Planktons :- 1. Phytoplankton	+	+	+	+	+	+	-	-	-	-
2. Zooplankton	+	+	+	+	+	+	-	-	-	-
Mosquito eggs and larvae of other animals	+	+	+	+	-	-	-	-	-	-

the test fish showed fast opercular movement and trying for come above the water surface.

DISCUSSION

Pollution is a serious problem ever since industrial effluents are disposed into water sources and on lands. It has increased with the growth of industry as well as population. Industries can conveniently be classified into two groups' viz., (i) Dry process industries and (ii) wet processes industries. Dry processes industries mostly engineering while wet process industries use water either as raw material or in their process or for both. Only a little quantity of water is absorbed in the process and the rest being discharged as effluent. Practically all the values of water used are calculated as mean values of effluent. In addition to the process of water large volumes of cooling water are also discharged as waste. Boiled below down is also an effluent contributing significant quantities to the bulk. These effluents contained a wide variety of materials of both organic and inorganic nature, including toxic substances and are usually discharged with or without treatment into surface water such rivers, this mixture some time gives colorful appearance to the effluent over.

The colour of effluent collected for study was found to be variable such as from wine red to reddish violet, yellowish black to dirty red, dirty yellow, dirty orange and white with yellow tinch, during different seasons at sampling sites, so present findings positively correlates with the reports of mentioned by earlier workers. Similar liquid industrial wastes also are of great concern because of their harmful effects (Lund, 1971) and Manivasakam (1984).

Industries use variety of raw materials chemicals which are later discharged via their effluent. Acids, alkalis toxic metals pesticides and other poisonous substances such as cyanide dyes, oils, detergents, resins, rubbers are a few to mention. Some of the effluents discharged tanning and metal pickling may contain pathogen (Bulton and Klein, 1971). The nature and extent of pollution depends on the materials present in the effluent and on the quantity discharged (Sawyer and M. Carty

1978). The component present in effluents when discharged to rivers in the form of industrial effluents. Organic acids which are discharged from the rayon, manufacturing, leather, chemical industries etc. in addition, nitric acid, hydrochloric acid and phosphoric acid are also discharged Nemerow (1953) and Masselli (1959).

Work of Kolkwitz and Marsson (1908) Bonznik, 1969; and Raza and Vijaykumari (1989) showed correlation with the organisms associated with different population zones and used as biological indicators of pollution in their saprobic system. A result of present study coincides with the above findings. The report also correlates with findings of Hynes (1960) when it was observed that the Physico-chemical parameter plays an important role in growth and development in aquatic animals.

In the present study when sample were collected had various types of smells such as wine, concentrated aromatic, unpleasant, rotten eggs likes unpleasant ammoniacal smell etc. Might be due to the receiving of variety of effluent at time from different sources.

Many workers have studies seasonal variation in floral and faunal abundances under different environmental condition. In the present study biological occurrence have been consider as an bio-indicator to know the pollution load at the site selected for investigation at Naregaon nullaha. Change in the water quality tends to change the biological environment which affects the growth fauna and flora, Rousefell and Everhart (1960). Quantity of phytoplankton undergoes a significant seasonal variation like other tropical water bodies, Tayyabsaify et al. (1986). The seasonal variation in diurnal movements of phytoplankton does not show any regular trend in their occurrence where as blue green algae show clear picture of variation with maximum quantity at surface (Singh 1990).

In the present study fluctuation in the occurrence of aquatic plant and other plankton development found very significantly. Green algae also found affected in the growth and abundance, these negative development might be due to pollution load in nullaha. These results correlate with earlier Reports of Rounsefell and Everhart (1960), Patil (1986), Tayyabsaify et al. (1986) and Singh (1990).

Zooplankton showed more distinct diurnal variation than phytoplankton. Seasonally fluctuation in diurnal variation did not show any significant similarity between two years. These observation correlates with the finding of Jolly (1952) and Mathew (1972). When the algal pollution was associated with photosynthetic activity through out the experimental site. The absences of diatom in effluent are due to the presence of certain pollutants (Venkateswarlu and Samapatkumar 1982). The maximum population of aquatic insect larvae and eggs are found in summer and monsoon month while the negative growth was reported in pre summer and post monsoon month. The negative growth may be due to heavy pollution load, unavailable of suitable foods over crowding and unfavorable density in dependent factors. Disappearance of aquatic animal plant are recorded at many of pleases of sites (Zoo and phytoplanktons) these observation correlate with the finding of Benke (1972 & 1979) and Rao (1980). Mohmad et al. (1986) and Trivedy and Goel (1986).

Rainy season, plays an important role in regulating various seasonal biological rhythm. It changes chemical concentration with influences the fluctuation in the quantity of phytoplanktons at maximum. Entire aquatic life depend directly or indirectly up on phytoplankton and governed by interaction of number of physical chemical and biological process (Singh 1990) and when nitrates and phosphates are at low concentration (Komarovsky 1953), Philipose 1959 and Carter 1960), where as the maximum production of diatom has been reported during winter season (Sawle 1964), Munwar 1970, Singh 1990) similar observation were recorded during the present study while analyzing the sample so the finding are coincides with the above quoted observation. It has been reported that free CO_2 can accelerate the growth of auglenophycae but in the summer the population decline as CO_2 level increase (Chakraborty et al., 1959, Jhingran 1983, Sharma 1984).

Table 18.4: Behavioural observation of test fish, *Rasbora daniconius* at different concentration of effluents (concentration in %)

EFFLUENT IN %	SAMPLING POINTS									
	1	2	3	4	5	6	7	8	9	10
100	Jerky and trying to come out of water	Fast moving	Fast opercular movement	Steady	Normal	Normal	Normal	Normal	Normal	Normal
50	Jerky with fast opercular movement	"	"	"	"	Normal	"	"	"	"
20	Fast movement steady opercular movement	No jerky movement	Normal opercular movement	"	Normal	"	"	"	"	"
10	Normal opercular movement	Steady swimming with normal opercular movement	"	Normal	"	"	"	"	"	"
0	Normal opercular movement	"	Normal	"	"	"	"	"	"	"

Table 18.5: Behavioural observation of test fish, *Channa gachua* at different concentration of effluents (concentration in %)

EFFLUENT IN %	SAMPLING POINTS									
	1	2	3	4	5	6	7	8	9	10
100	Jerky movement fast opercular movement and tries to come out of water surface	Fast opercular movement	Jerky but normal movement	Normal	Normal	Normal	Normal	Normal	Normal	Normal
50	Normal opercular movement	Normal	Normal		"		"			"
20	Slow movement Normal opercular movement		"		"		"	"	"	"
10	Normal opercular movement	Normal opercular movement	"	"	"	"	"	"	"	"
0	Normal opercular movement			"	"	"		"	"	"

There is an increase in abundances of a few species but range of individual species widens, (Bonzik 1969) has shown in his studies on population structure of phytoplankton in polluted water, the diluted distillery effluent can be rich source of plant nutrient and potentially to be utilized for irrigational practices Raza and Vijaykumari (1989).

Fishes were found absent in the water of Nullaha. This may be due to pollution which changes the biological environment either by destruction of food or change of the Physico-chemical character of water which affect fish fauna as stated by Rousefell and Everthart (1960). The results reveal that the qualities of phytoplankton undergo a significant seasonal fluctuation like other tropical bodies. Seasonal variation in diurnal movements of phytoplankton did not show any regular trend. Singh (1987) reported that the blue green algae showed a clear picture of diurnal variation with the maximum quality at surface water, it is further advocated that algal population was high found associated with photosynthetic activity similar result were obtained during the present study which shows correlation with the earlier report recorded by Venkateswarlu and Sampatkumar (1982). The maximum net production of aquatic population was found in summer and monsoon months. The negative growth may be due to heavy mortality of population, unavailability of their suitable food, over crowding and unfavorable density independent factors (Singh, 1982).

It has been observed that the amount of rain fall plays an important role in regulating in various seasonal and biological rhythms. The entry of rain water in to nullaha, changes the chemical concentration which influences the fluctuation in the quality of phytoplanktons. Maximum abundance show during the period of march-April. The maximum population of blue green algae higher during the period of November to June. When nitrates and phosphates were low. Maximum production of diatoms was observed during winter months.

Euglenophycea in the nullaha was rich during the period when free carbon dioxide was high. This indicate that Euglenophycea prefer free carbon dioxide for their growth.

Population of Euglenophycea decline during warmer month as compare to other month during both the years. Diluted distillery effluent can be rich source of plant nutrients and the potential be utilized for irrigation practices.

Nitrates enhance the production of plankton. Low vales of nitrates during maximum production can attribute to its consumption in nutrients by plankton directly as well as in directly.

Biomass production and phosphate go side by side i.e. increase in phosphates also enhance the production were as its low value are responsible for low production. Thus nitrates phosphates being of nutrients element and their availability in enhance the planktons production. Higher value of biomass noticed during high contain of DO, while negative value were at BOD. Salinity was found to be influence the production preferable during low salinity which corresponds to low pH and in turn higher dissolution of oxygen.

Summer is the only season when maximum growth of phytoplankton occurs simultaneously from the last week of April to whole of May. This condition remains almost same in January. The population of phytoplankton fluctuation with the decrease in water level and on set of winter as the monsoon commences no. of phytoplankton species rapidly falls, possible due to dilution but, by the end of July their production tremendously reaches maximum.

Zooplankton showed the more diurnal variation than phytoplankton. Seasonal fluctuations in diurnal variation did no show any significant change during the study. In the faunal study qualitative estimation of zooplankton and phytoplanktons found maximum in the rainy season more particularly the occurrence of zooplankton are maximum during monsoon season, it may be due to favourable condition of the rain fall and availability of more quality of food.

Behaviour is the "whole complex of observable, recordable, or measurable activity of a living animal" Verplank 1957, it is everything animal does including all of it integrated movement. Most animal behavioural patterns (responses) involve over movement that they are highly to environment viable (or stimuli) of a physical, chemical or biological nature. Though its normal adaptive behavior in animal tend to mitigates or bring itself in to favourable environment perturbation. Ultimately most favourable relationship is those that prompts the survival growth, reproduction and longevity of population

CONCLUSION

The present study it can be concluded that, the data on flora and faunal abundances and behavioural changes of test fishes in sample collected from different point is suppose to be the basic information to know pollution load of the reservoir the findings in the present study correlate to the reports sated earlier workers in the concern aspect. On the basis of present study it has been explicitly inferred that change in behaviour can serve as indicator of pollution. Fishes were found absent in the water of Naregaon nullaha. They may be due to unfavorable condition occur due to pollution which change the biological environment either by destruction of fish food or change of there type of organism which affect floral and fauna.

ACKNOWLEDGEMENT

Authors are grateful to, Head, Department of Zoology, Dr. Babasaheb Ambedkar Marathwada University, Aurangabad (Maharashtra) for laboratory facilities and Director of Board of College and University Development for financial support.

REFERENCES

Ajmal M. Azhar, A.N. Majahid A.K. (1982): Quality of Ganga river in Uttar Pradesh and Bihar IAWPC. Tech. Annual, 11:165.

APHA: 1986: Standard methods for the examination of water and west water 16th Edition.

Benke A.C. (1992) An experimental field study on the ecology of co-existing larval odonates, Ph.D. dissertation. Anthenes Georgia, Univ. of Georgia, 122 pp.

Boznick, E.G. (1969): Laboratory and field studies of phytoplankton studies. Ph.D. thesis, Washington University, Missouri.

Bulton, R. and Klein, L. (1971): Sewage Treatment Basic Principles and Trends, Butter worths London.

Carter, C.S. (1960): Nature, 189: 843.

Chakraborty, R.D., Ray, P. and Singh, S.B. (1959): Indian J. Fish, 6 (1):186.

Hynes, H.B.N. (1960): The biology of pollution water. Liverpool Univ. Press, London, 621.

Jhingran, V.G. (1983): fish and fisheries of India. Hindustan Publication Corporation India, Delhi.

Jolly V.M. (1952): Hydrobiologia, 25: 466.

Kolkwitz, R. and Marsson, M. (1908): Okolagieder Ptlanzlichen saprobein. Ber. Dentsch, Bot. Gen., 26: 505-579.

Komarovsky, B. (1953): Bull. Res. Counc. 2 (4): 379.

Lund, H.F. (1971): Industrial Pollution Control Hand Book. MgGraw-Hill Book, Co., New York, 158-168.

Manivasakam, N. (1984): Physico-chemical Examination of water, sewage and Industrial Effluents, Pragati Prakashan, Meerut, 268-292.

Masselli, (1959): "A simplification of Textile Waste Survey and Treatment New England Interstate Water pollution control commission.

Mathew P.M. (1972): Ph.D. thesis, Agra University, Agra., India.

Mohammed A., Habibullah C.M. and Qadri S.S (1986): Fresh water zooplankton fauna of Mir Alam Lake, Hyderabad, Andra Pradesh and their estimation, qualitatively and quantitatively with reference to the season. Jour. Sci. Res. 8 (2 & 30) : 63-66

Munawar, M. (1970): Hydrobiologia 36 (1) : 105.

Nemerow N.L. (1953): Oxidation of cotton Kier wastes". Jour. of Sewage and Industrial Waste, 88-92.

Patil S.G., Singh, D.FR. and Harshey, D.K. (1986): Impact of Gelatine factory Effluent on the water Quality and Biota of a stream Near Jabalpur, M.P. J. Environ. Biol. (1): 61-65.

Phillipose, M.T. (1959): Proc. Symp. Algol, I.C.A.R., New Delhi, p. 272.

Rao, M.N. and Dutta A.K. (1979): "Waste water Treatment" (ed). Oxford and IBH Publication Co. Calcutta. 320-32'6.

Raza S.H. and Vijayakumari N. (1989): Impact of distillery effluent spent wash, on seed germination morphological characters yield land pigment concentration of trigonella foenum, Graecum. Poll. Res. 8(3): 109-116.

Rounsefell A. and Everhart W.H. (1960): Fishery science its methods and application. John Wiley and Son Inc. New York, London. 168-192.

Sawle E.M.F. (1964): J. Ecol., 52:433.

Sawyer, C.N. and M. Carly, R.L. (1978): Chemistry for Environmental Engineering. 56-62.

Sharma R.C. (1984): Potamological studies on latic environment of the upland river, Bhagirathi of Garhawal Himalaya. Environment and Ecology, 2: 239-241.

Singh H.R., Veena Baduni and Manoj Negi (1987): Phytoplankton as an Aquatic Resource. Abstr. Proc. 8th AEB SEJ and Symp.Jammu. (21).

Venkateswarlu V. and Sampatkumar, P.T. (1982): Chemical and biological assessment of pollution in the river Moosi, Hyderabad, (A.P.), India, Proc. Acad. Environ. Boil. 1 (1): 89-94.

Singh, D.N. (1990) : Diurnal vertical migration of Plankton in McPherson Lake, Allahabad. Proc. Nat. Acad. Sci. India, 60 (B): II pp. 141-152.

Singh, D.N. (1990): Studies on the Phytoplankton of Mc. Pherson Lake, Allahabad. Proc. Acad. Scio. India, 60 (B) : III:pp. 237-243.

Singh, D.K. and Singh, C.P. (1990): Pollution studies on river Subernarekha Around Industrial Belt of Ranchi (Bihar).

Somushekar, R.K. (1985): Studies on water pollution of river Cauveri. Bull. Bot. Sco. Sagar, 32 : 145 - 149.

Tayyabsaify, S.A., Chaghtai, Praveen Alvi and Durrani, I.A. (1986): Hydrology and Periodicity of Phytoplankton in the sewage fed Motia pond Bhopal, India. Geobios, 13 : 199-203.

Trivedi, R.K. and Goel P.K. (1986): Characterization of treatment and disposal of waste Water in a textile industry. Ind. Poll. Cont, 2 (1) : 1-12.

Verplank, W.S. (1957) : A glossary of some terms used in the objective science of behaviour. Phychol. Rev., 1-42.

19

Discarded Medicines and Cytotoxic Drugs

S.G. Yeragi

INTRODUCTION

Little drops do an ocean make. If every household take care of the microenvironment in an around their dwellings the environment as a whole will improve and also the quality of life.

The house holders should know how to handle and dispose the wastes without causing an adverse impact on the immediate environment around. The vegetable and fruit pilings and discarded portion of vegetables can be fed to street cows or to the birds. We will find Neem plants sprouting in our compound. The spent tea powder and onion peels can be used as manure for the rose plants. Our home is in fact a micro-eco-system.

Relation of man with the atmosphere is necessarily symbolic. It is necessary to maintains proper equilibrium between two must be at all costs. Last few decades man's relationship with the atmosphere has drastically changed due to vast increase in his expectation and activities.

Today ignorantly man can add all the waste materials in the atmosphere, which affect the oxygen levels and hence population explosion demands are made on the limited sources of energy and materials. The capacity of the environment to support the increasing demands of man is limited. Environment performs three basic functions in relation to man, provides

K. J. Somaiya College of Sc. and Com. Vidyvihar, Mumbai - 400 077.

living space and other amenities that make life qualitatively rich for man. Source of agricultural, minerals, water and others, consumed directly or indirectly. Environment is a sink whose all the waste produced by man is assimilated. Human population human explosion, an impurity of air is also increasing side. Indian environment contain less amount of oxygen due to several reason like BOD, COD etc. Healthy man inhales about 16.5 kg Air/day and child double of this. Due to deficiency of oxygen in air, man has to inhale more and more air to get enough oxygen for respiratory metabolism but intake of particulates load increase in his lungs which help to disturb the physiology causing allergy, bronchitis, asthma by taking treatment with consuming medicines and contaminating the environment. As you pollute the environment more and more then the meditation also directly correlated to you, causing complications of the health. Everyone should remember that breath good way by consuming enough oxygen from the nature.

Are discarded drugs contaminating the environment?

Certain chemicals in drugs and some foods are now detectable levels in the environment. In the survey in all states found minute amounts of dozens of antibiotics, hormones, pain relievers, cough suppressant, disinfectants and other products. Some of this is due to the disposal of unused prescription drugs in hospitals and homes. Multi-therapy helps to increase the bulk of medicines and after the cure; no one can bother about how many drugs are stored. During festival time under the head of cleaning of home, all expire date medicines are randomly dump in the nature. No one is bother or thinking about their adverse effect on others. Human has the habit to use, cure and then through anywhere in the environment. After cured any dreadful diseases no none has the habit to take care of left or expiry dated medicines. We can't regulate how a consumer disposes of something. We can only provide advice. In conjunction with the new regulations, they plan to issue "Best practices" documents aimed at consumers, manufacturers and water monitoring agencies and perhaps hospitals. They will also urge physicians to reduce the number of prescriptions they write.

New products submitted for approval under the food and drugs act will have to undergo an environment assessment under the new substances Notification, Regulations of the Indian environmental protection act-US food and drug administration is also considering requiring more tests to determine the effect of drugs on the environment.

The use of chemotherapy as a cancer treatment modality is increasing. Cytotoxic chemotherapy may be used for non-malignant conditions. These guideline should be followed wherever cytotoxic chemotherapy is administered. Medicines used in cancer chemotherapy may be carcinogenic, mutagenic and teratogenic.

Aim of guideline is to assist health care processionals to:

- Recognize the risks and cause of exposure.
- Minimize exposure.
- Take appropriate action if exposure occurs.

General principles for handling cytotoxic chemotherapy

- Staff working with carcinogenic agents must be made and kept aware of risks and the circumstances under which they play be exposed to the carcinogen. This guide must be available to all staff involved with the preparation administration and disposal of cytotoxic chemotherapy.
- Manager must ensure that appropriate training and assessment of practice is undertaken so that staffs are competent in the handling of cytotoxic chemotherapy.
- Only trained staff should allow to carry out the work.
- Currently cytogenic examining cell change and analytical methods for measuring levels of cytotoxic chemotherapy in blood and urine are available.

Work is ongoing to develop

- Procedures for measuring levels of cytotoxic in pharmacy and clinical areas this will allow.
- Staff competency testing
- Evaluation of new procedures and technique.
- Validation of cleaning procedures.

Few names of the drugs still marketed in India but banned by Governments abroad due to serious side effects as per WHO as Analgin, Droperidol, Furazolidone, Phenformin, Phenolphthalein, Piperazine.

Pregnancy and Breast-feeding in staff handling cytotoxic chemotherapy

- Complex and emotive issues which are at extreme importance to staff in areas predominately staffed by women of child bearing age.
- Various studies have shown links between occupational exposure to cytotoxic chemotherapy and miscarriage, menstrual dysfunction and infertility.
- Staff who may be pregnant should refer themselves to the Trust occupational health service for advice before deciding whether or not to continue handling cytotoxic chemotherapy.
- Pregnant, breast feeding or carrying for babies under 26 weeks old are considered to be at risk.

Disposal of equipment:

Unused solution

- Any cytotoxic solutions remaining in the ampoule or vial at the end of the preparation procedure must be drawn up into a syringe for checking and disposed of by incineration at 1000°C.
- Any expired or unused cytotoxic preparations supplied to clinical areas must be returned to the pharmacy in an appropriate cytotoxic drug transport container for safe disposal.
- All other materials like masks, gloves, disposable trays, towels, dressings etc. are placed in a polythene bag, tied and then placed in the recognized yellow clinical waste bag ready for disposal in the incinerator. Safety glasses must be washed carefully with soap and water. The operator must be washed hands thoroughly at the end of the procedure.
- Tablets or capsules present a lower risk than solutions while blister or foil packaging provides further protection.
- Preparations should be labeled with a warning that any person other than the patient for whom the drug is prescribed should avoid touching tablets or liquids directly.

Cytotoxic chemotherapy

- Actinomycin-D, Amsacrine, Dounorubicin, Epirubicin, Vinblastine, Vincristine, Vindesine, Vinorelbine.

Glossary

- *Carcinogen (ic)*: A cancer causing agent
- *Extravasation:* The leakage of a medicine into the substances tissue that is capable of causing pain, necrosis and/or sloughing of tissue,
- *Flare:* A local allergic reaction without pain but is usually accompanied with red blotches along the vein subsiding within 30 minutes with or without treatment.
- *Irritant:* An agent capable of causing aching, tightness and phlebitis at the injection site and/or along the vein with or without an inflammatory reaction.
- *Mutagen (ic):* An agent with the capacity for causing genetic mutations or chromosomal aberrations.
- *Teratogen (ic):* Anything capable of causing or forming a blister and/or tissue destruction.
- *Medicine:* It is the solution to cure the dreadful diseases. One disease and multiple therapy take time to worn the pathogen. Medicines either killed the pathogens or neutralize their toxicity to get relief to the patient. There is an adverse effect of medicinal therapy for example whenever there is headache, immediately allopathic medicine like Anacin, Crocin give quick relief but affect the nervous system causing probably leukemia. The pain killer drugs are dangerous. Generally experience doctor avoid giving antibiotic only because of adverse effect and exercising for the lymphocytes to produce antibodies and antitoxin. Multiple therapy gather lot of drugs which ultimately thrown in the sink of environment. Man is consuming less drugs but discard more in nature. In the nature, such dreadful substances entered in the food chain and finally get way to entered and stored in human body by the process called bio-magnifications. Expiry date of the drug cause the toxic effect to human being due to introduction of intra chemical substances. Expiry makes the chemical constituents free and toxic. The discarded drug severely affects the

environment. Mina mata diseases, Ita-ita diseases are due to heavy metals. In West Bengal many people are under the treatment of knee joint pain because of iron mining discharge heavy dose of Iron in drinking water. Hopeful is common pain killer affect breathing, aggravate drug toxic to human. Aniniaherin is photosensitive drug affect eye sight. Ayurvedic drugs are therefore taken before sunrise or after sunset. Old elderly people are always using wooden umbrella to avoid the effect of uv rays. Tetracycline is an antibiotic sensitive to light and causes skin problem like leukorderma, the white patches on the skin. Metoniadzole is given to the patient in India but ban abroad. It causes liver cancer. Midxalic is antidyrial but photosensitive drug.

American women use to take tetracycline for the safety of baby but baby born will yellow teeth called fluorosis. The child suit the case against mother. The adverse effects of the dreadful drugs are as they are mutagenic, carcinogenic, teretogenic, vesicant, irritant etc. The heavy metal blocked the replication of DNA causing abnormality in the children.

Tuberculosis

There are several drugs used in TB like Antituberculosis, Isoniazid, Ehambutol, Pyrazinamide etc. These drugs mostly cure the disease but directly affect the optic nerve.

In developing countries tuberculosis is still a major public health problem both in children as well as in adults. In 1956 there were fifteen lakhs patients in India. Now it is estimated that there are about 14 millions TB patients in our country of whom about 3.5 million are infectious (sputum positive). About 2.2 million TB patients are added every year out of which one million are sputum positive. In India five lakh people die due to TB every year i.e. one patient in every minute so more than one thousand people die every day. It is necessary for further intensive research is to be done.

In 1929 there was only one antibiotic named Penicillin but today there are 7000 antibiotics. In US every year 35000 new drugs are investigated only on leukemia so in what way there are discarding such huge medicines.

Alternatives to the dreadful drugs

No body is attracting toward ayurveda and their alternatives. There are several alternatives like maggot therapy, leech therapy, Ram Dev medicinal therapy, Morarjibhai therapy, Acupressure, Yoga etc.

Toxicity mostly reduced by heating or incineration, so discarded tablet should not open or crushed but heat so that their toxic level will come down. Cobra venom when treated above 60°C the neurotoxin gradually goes on reducing. Shankara's neck had cobra because of cardiotoxin present in venom which accelerate cardiac output therefore of the serious myocardiac arrest the cardiotoxin injection given to the patient.

Recycling of drug is another way to save the effect of drug on nature. If heat killed medicines are given to the plants i.e. phytoremidiation or bioremediation help to save the nature. To stop producing varieties of drugs on one diseases, the chemical formula goes on changing by labeling new drug. This should stop in the favour of environment.

Fluoride

WHO has allowed the permissible limits of fluoride in drinking water 1.0 mg to 1.5 mg/L. Health point of view, fluoride is bit an essential element as well as toxic one. It helps in the formation of dental enamel and mineralization of skeletons especially in young ones. Macro amount causes discoloration of teeth enamel (dental fluorosis). Fluoride is chemically the highly reactive electronegative ion and widely distributed elements in the earth's crusts. It is easily soluble in water. In south Konkan well water contain high level of fluoride hence more dental fluorosis is found in coastal living natives.

Neem a grand tree is insecticidal antipollutionary effect of Neem (Azadirachta), Tulsi (Oscimum) and their role in purification of air is well known in Indian people especially in villages. 1980 international Neem conference was organized in Germany.

CONCLUSION

The unplanned use of natural resources like land, water, forests, fuels, mining etc. causes imbalance of ecosystem. The man is the main pollutant who used these sources continuously throwing the wastes in nature. Ocean has 97% water of the tidal, human can not used directly due to salinity, 3% fresh water of which 2% polar snow and remaining 1% of water is on disposal to meet our needs.

- Let us breathe more oxygen than CO, SO_x and NO_x
- Let us not waste our money on medicinal check up and medicines.
- Let us conserve fuel and reduce foreign exchange out flow.
- Let us cut down fuel cost on our vehicles to save money.

Are you aware that

More than 40,000 people are dying prematurely every year in 36 Indian cities because of polluted air.

As per natural selection, nature used to select good one and rejected weaker one so that defective genes naturally omitted and selected and inherited good one to produce healthy progeny. Now-a-days human being for survival pint of view taking medicinal and undergo gene therapy so but natural drugs are continuously added and discarded dump in the sink of the environment. These drugs are distributing the food chain. DDT is non degradable insecticide totally banned in the world but still used in India. Its toxicity remain in the nature for hundred of years as it is and make chromosomal defect causing abnormality in newly born children. Darwin's natural selection concepts are still going on in environment. As a matter of fact the Gia theory of Jomes Lovelook, which considers the earth as a gigantic living organisms exemplifies the ideal economic concept. According to Lewis the earth is a living cell and atmosphere is its protective covering which andropogenic activities are destroying using environment as sink.

ENVIRONMENTAL ETHICS

Basic scientists define environment as "Aspects of physics, and chemistry of air, water and soil, where living thing live." But when it comes to human the definition also includes "social, cultural, political and economic aspects". The ecocentric and biocentric ethics weigh heavily on the earth and living system as a whole and emphasize a symbiotic relationship "Vasudeba Kutumbakum". All equations in these systems function under" a rule that everything is connected to each other things for mutual benefit and sustenance.

Ultrastructural Changes in Splenic Cells and Intermediate Vascular Pathway of *Mus musculus* under Sub-Lethal Dose of Endosulfan

*Preety Sinha, Reena Sinha and A. Nath**

SUMMARY

In the present investigation, endosulfan, a highly toxic compound, was selected to observe its deleterious impact of spleen of adult female Swiss albino mice, *Mus musculus*. They were administered the oral dose of endosulfan (3 mg/kg bw) per day for four consecutive weeks. Treated and control specimens of mice were sacrificed on the 30th day. The tissue of spleen was isolated from control and endosulfan exposed mice. Tissues were studied under light and electron microscopy. The light microscopy revealed various damages and distortions to splenic cells because of toxicity caused by pesticide induced. The spleen was intensively congested, looking considerably enlarged with dark brown or red colour in appearance. The capsule of the organ was thickened and was adherent to the neighbouring tissues. The venous sinus (red pulp) was dilated and the endothelial cell lining them was prominent. The walls of sinusoids were thicker and more fibrous than normal. Degeneration of splenic nodules was seen. There was complete distortion of white pulp. Increased number of macrophages was also seen. Degenerative stages of intervascular pathway

Department of Zoology, A.N. College, Patna,

*Department of Zoology, Patna University, Patna

were observed. Under electron microscope, distorted mitochondria, increased endoplasmic reticulum and number of osmiophilic granules were seen. These are the significant signs of increased toxicity.

Other abnormalities and the degree of impairment in the splenic cells, increased with the increase in concentration and duration of endosulfan administration.

Key words : Swiss albino mice, spleen and endosulfan.

INTRODUCTION

Endosulfan is a chlorinated cyclodiene compound, used worldwide as an insecticide and agaricide. Its increased and indiscriminate use in agricultural practices has greatly enhanced the chances of its exposure to animals and human beings. It is an important cause of pesticide poisoning in many countries including India. Several studies have clearly established the harmful effects of various hazardous pesticides on animals. Spleen is the most important lymphatic organ as it makes disease fighting components of immune system and also removes abnormal cells after filtering blood. There are some reports of harmful effects of pesticides on splenic cells.

Wojdani et al. (1984) have studied antigen sensitized splenic lymphocytes and peritoneal exudates lymphocytes. They have concluded that NK cell activity appears to be irreversibly inhibited by polycyclic aromatic hydrocarbons carcinogens. A significant increase in splenic natural killer cell activity in female mice was reported by Crittonden et al. (1998) upon treatment of methyl parathion. Rai (2002) studied the effect of dimethoate an organophosphorus compound on spleen of mice. Kostka at el. (1995) and Russel (1995) studied the effect of organochlorine insecticide on liver and testicular cells of mice.

Only a few records of work are available on the effect of Endosulfan, an extremely hazardous pesticide on spleen of mice. Therefore, an attempt has been made to observe the ultrastructural changes in splenic cells of *Mus musculus* after exposure to endosulfan.

MATERIALS AND METHODS

In the present investigation, experiments were performed on 3 to 4 months old healthy female Swiss albino mice (*Mus musculus)*. Thirty fresh and healthy mice were taken and grouped randomly into three different cages with specific codes.

Ten of them formed normal group, other ten as the control group and the rest ten were selected for treatment by endosulfan. Mice of all the groups were provided all facilities on the same scale.

The LD_{50} value of endosulfan for mice was calculated by standard interpolation method and extoxnet standard data reference which was 7.8 mg/kg body weight.

The sub-lethal dose of endosulfan (EC 35%, obtained from local market, manufactured by Excel Industries LTD, Mumbai, India) was dissolved in distilled water to make the dose of 3 mg/kg bw. The selected dose was administered daily to the experimental mice for four consecutive weeks.

Normal control and treated mice were sacrificed on the thirtieth day of treatment. Spleen of different groups were dissected out and fixed for light and electron microscopic study.

For light microscopic study, tissues were fixed in buffered formalin and Bouins fixative processed. Tissues were processed, paraffin blocks were prepared and sections were stained with haemotoxylin, eosin and mounted in DPX.

For electron microscopic study, tissues were fixed in Gluteraldehyde. After preparation of ultra thin sections microphotography was done.

RESULTS AND DISCUSSION

Observations under light and electron microscopy of normal and control group show same histological structure. No changes were observed in both group. Under light microscopy treated mice reveals damages and distortion in splenic cells caused by toxicity of induced endosulfan. Spleen of treated mice was intensively congested. It looked considerably enlarged with dark brown or red colour in appearance. Capsule

OBSERVATIONS
PLATE - IA
Light Microscopy

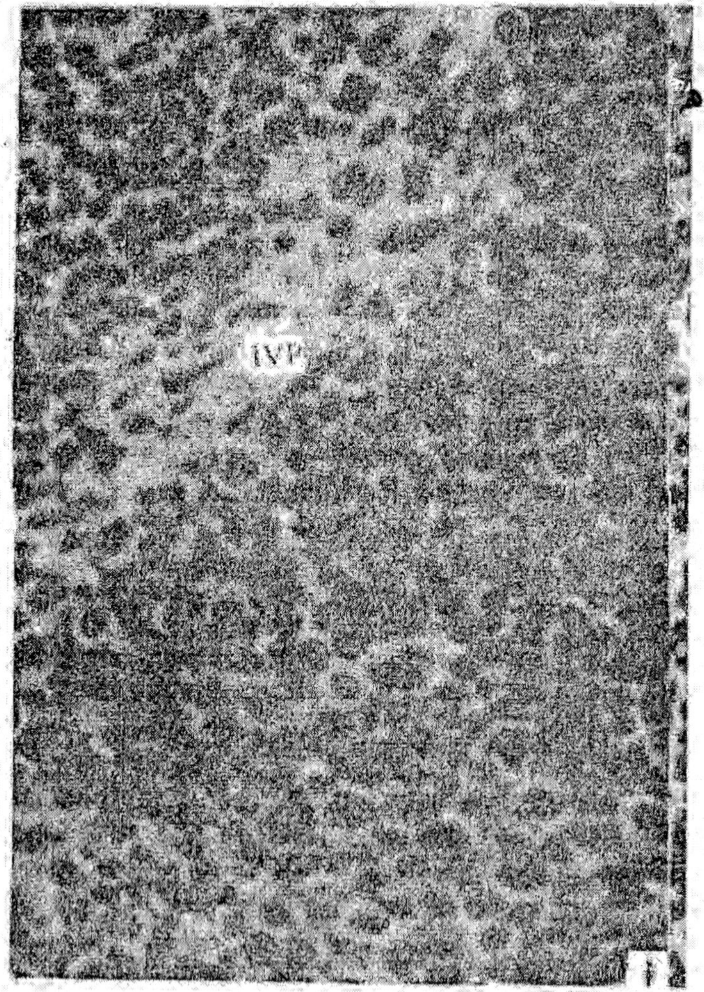

Fig. 1 : Spleen of control *Mus musculusL:* Intervascular pathway is significant. A meshwork of reticular cells was also significant. Different stages of formation of erymrocytes were also prominent. X 450.

PLATE - IA

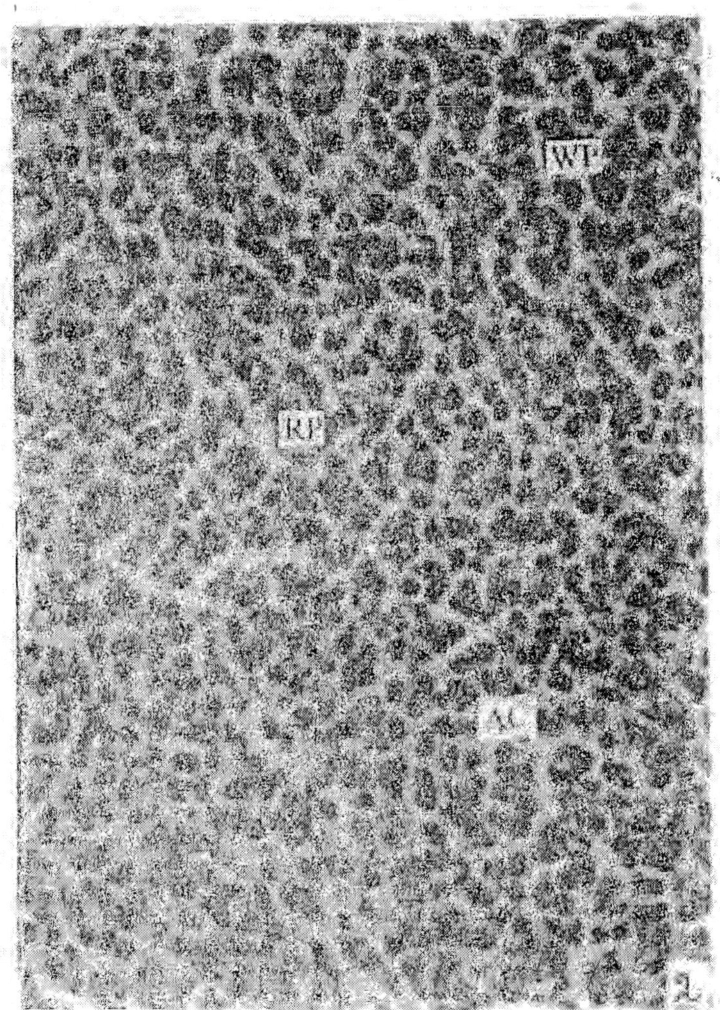

Fig. 20.2 : Spleen of treated *Mus musculus*: Changes in organization of reticular cells were observed, shrunken splenic cells with condense nuclei showing beginning of apoptosis and formation of apoptotic cells (AC) can be seen. X 450. Red pulp (RP) and White pulp (WP) were not very distinguishable.

PLATE - IB

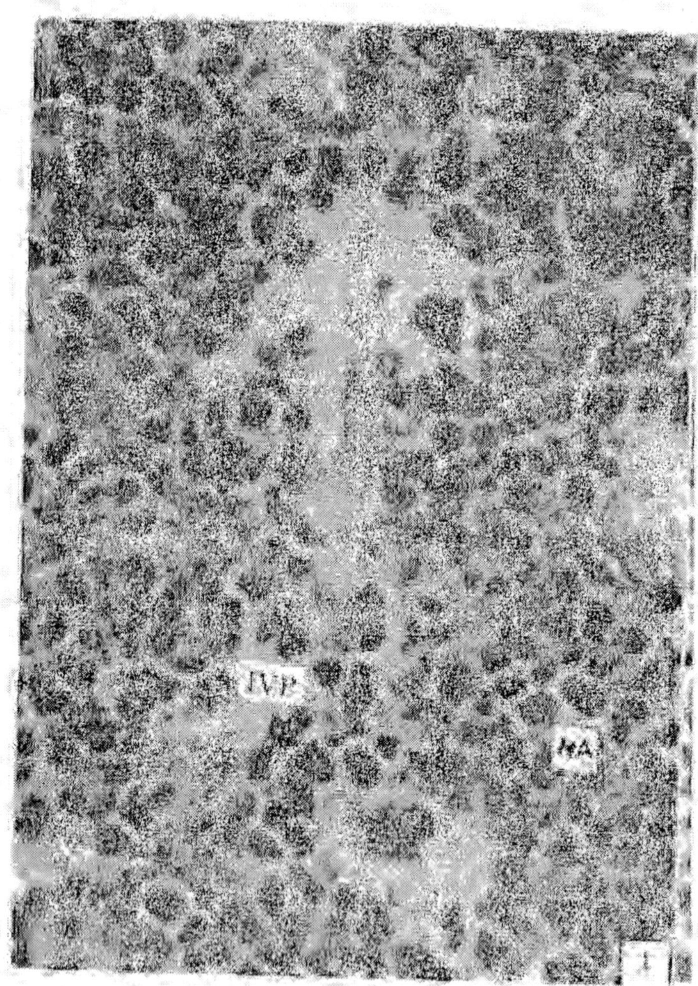

Fig. 3 : Spleen of endosulfan treated *Mus musculus*: Degeneration in splenic nodules and advance stage of intervascular pathway were seen. There were clumping of RBC in arterioles and veins. Beginning of the scattering of macrophages was well observed. Small dark blue apoptotic cells were observed X 450.

PLATE - IB

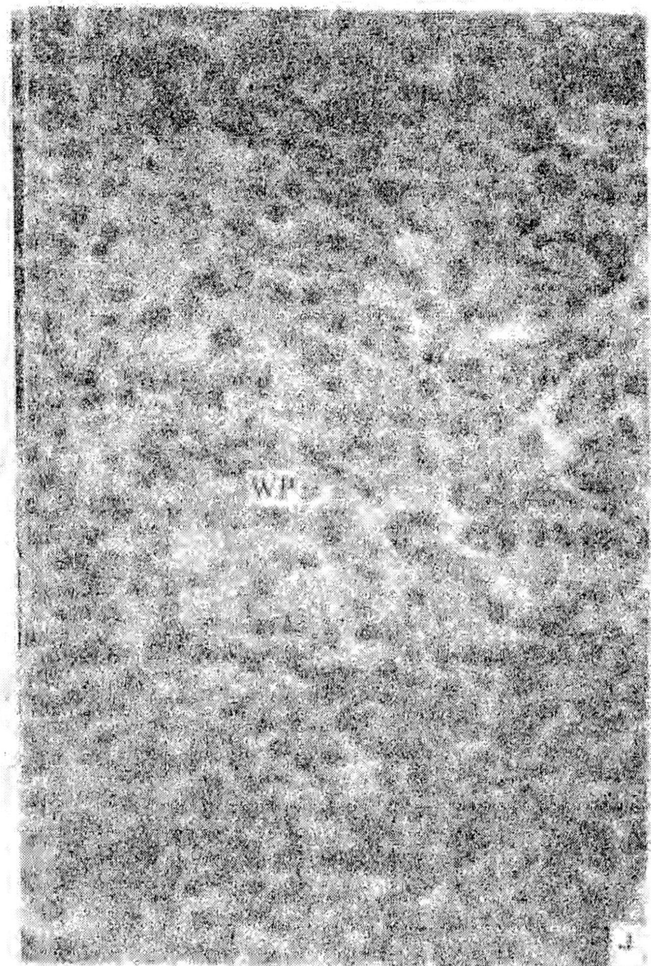

Fig. 4 : Spleen of endosulfan treated *Mus musculus:* Complete distortion of white pulp. Increased number of macrophages was prominent. Complete bursting of red pulp and oozing out of red cells were clearly observed. Small compact shrunken dark blue nuclei were significant which were signs of apoptosis X 450.

IVP – Inter vascular pathway; WP – White pulp; MA – Macrophage.

PLATE-II

Electron Microscopy

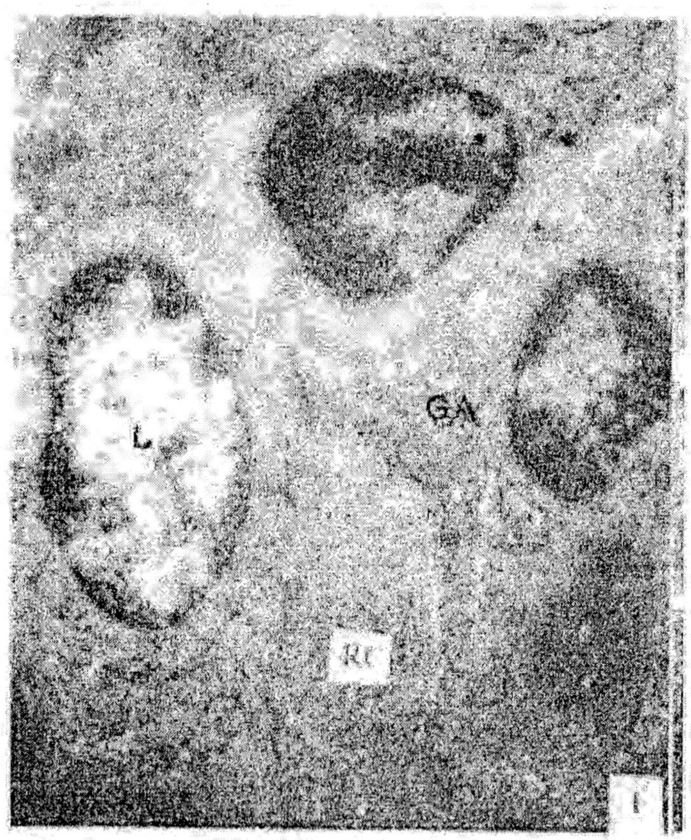

Fig. 1 : Spleen of control Mus *musculus:* Lymphocytes with large number of ribosomes through out the cytoplasmic matrix were seen. Golgi area (GA) was relatively small and consequently inconspicuous. Endoplasmic reticulum was rare. The reticular cells (RC) were distinguished from lymphocytes by relatively small amount of condensed chromatin materials. The cytoplasm of the reticular cell (RC) resembles fibroblast X13000.

PLATE-II

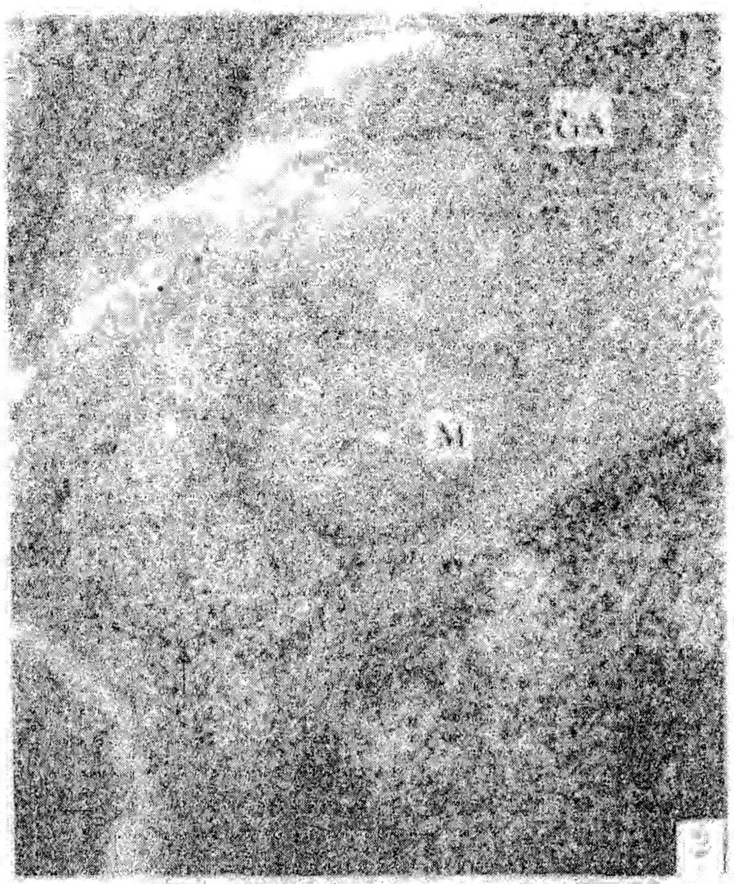

Fig. 2 : Spleen of control *Mus musculus:* The inner and outer membranes of the Mitochondria (M) were well confirmed within the mitochondrial matrix, cristae are present. Ferritin granules were present (↑) X 90000.

PLATE - IIIA

Electron Microscopy

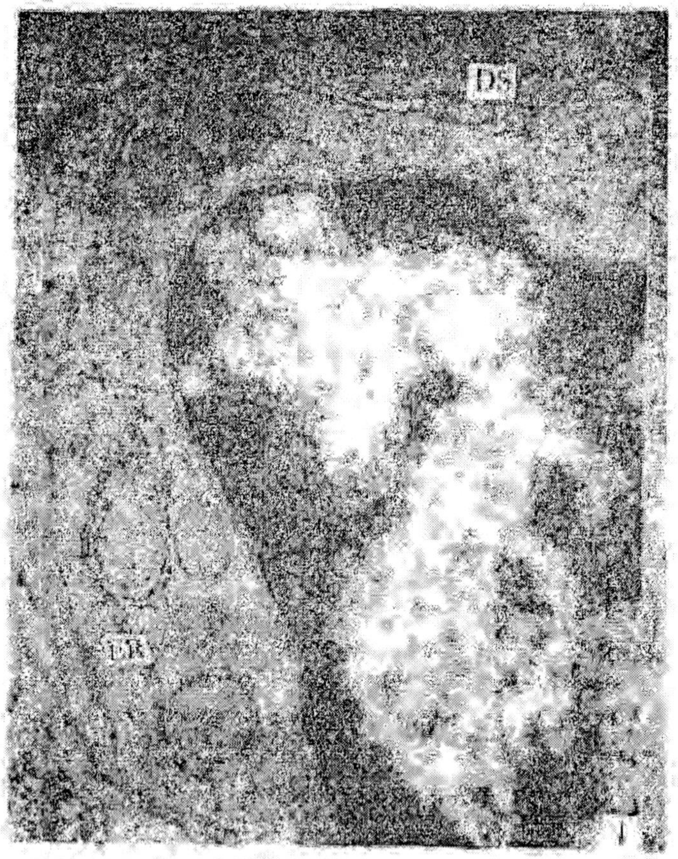

Fig. 1 : Spleen of endosulfan treated *Mus musculus:* Peripheral condensation of chromatin material. The endoplasmic reticulum (ER) around the two mitochondria (M) was significant. Dissolution of mitochondrial cristae. Serrated margin of plasma membrane and bulging out of plasma membrane at desmosomes (DS). X 75,000.

PLATE - IIIA

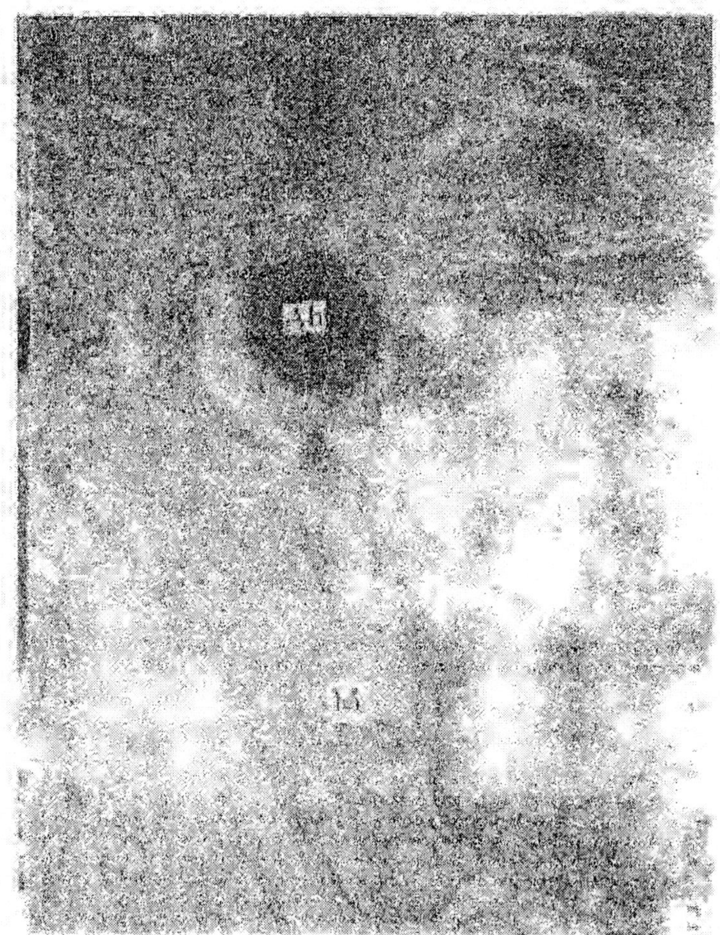

Fig. 2 : Spleen of endosulfan treated *Mus musculus:* Lymphocytes showing phagocytotic process of aggregating apoptotic body (Ab). The dissolution of cristae and inner membrane of mitochondria were seen but outer membrane were least affected. Nuclear membrane was composed as seen in normal. X 75,000.

PLATE - IIIB

Electron Microscopy

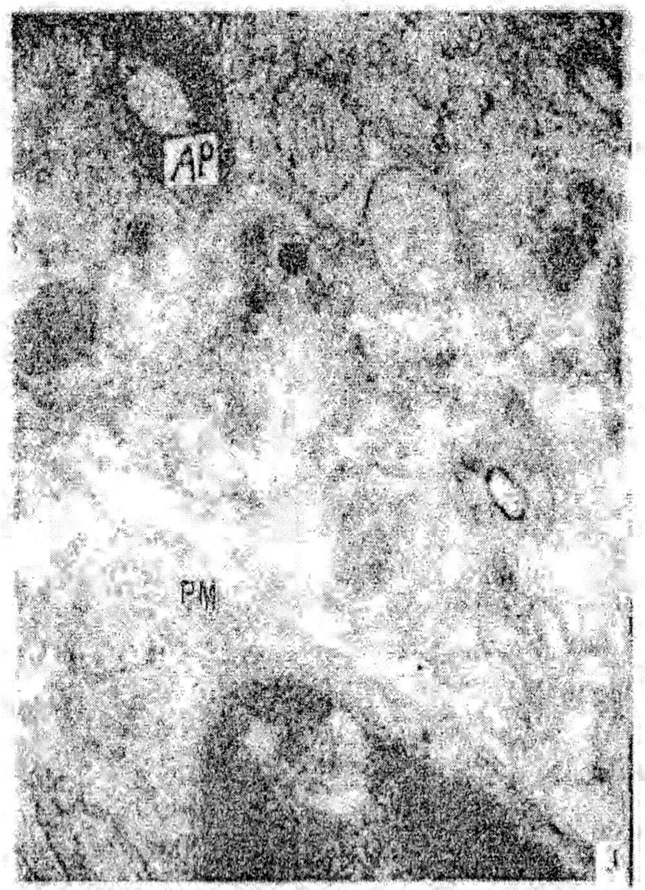

Fig. 3 : Spleen of endosolfan treated *Mus musculus;* Fusion of two cells were seen. Dissolution of plasma membrane (PM) at few places was noted. Formation of autophagosomes (AP) is very prominent, Lysosomes engulfing mitochondria. X 75,000.

PLATE - IIIB

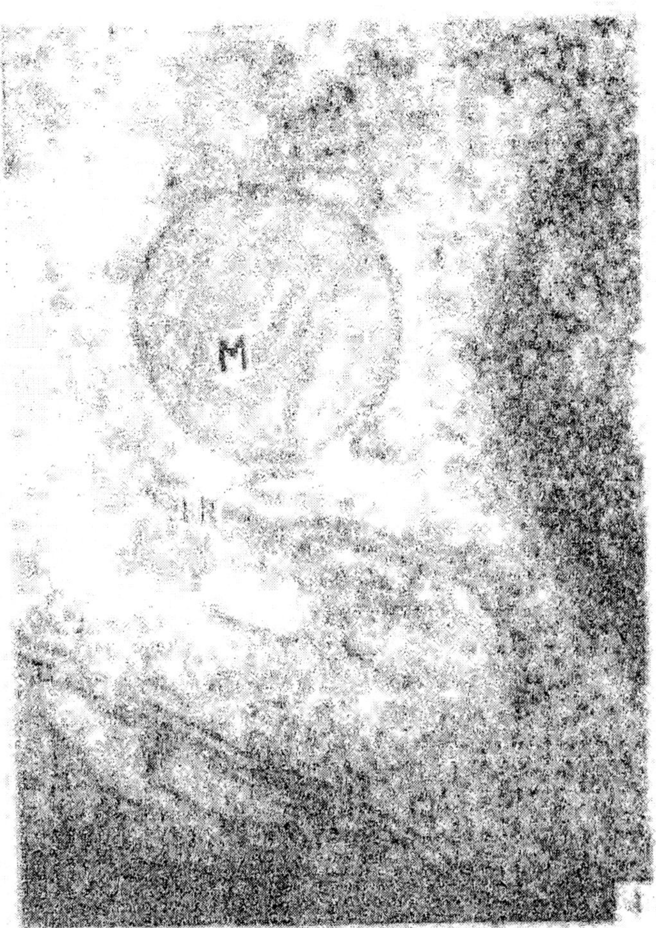

Fig. 4 : Spleen of endosolfan treated *Mus musculus:* Formation of arch like structure around the mitochondria (M). Presence of condensation of osmiophilic granules. The outer membrane of the mitochondria is some how looking serrated. Cell organelles are scattered and never show tendency of specific zones. Endoplasmic reticulum (ER) was rare. X 75,000.

of the organ was thickened and adherent to the neighboring tissue. The red pulp (venous sinus) was dilated with prominent endothelial cell lining them. The walls of sinusoids were thicker and more fibrous than normal. There were evidences of degeneration of splenic nodules. White pulp showed complete degeneration. Increased number of macrophages was also seen. Intervascular pathway showed degenerative stages.

Under electron microscope, mitochondria showed degeneration. There were evidences of increase in endoplasmic reticulum and also in number of osmiophilic granules. There were significant signs of increased toxicity, which ultimately led to nuclear damages. This explains why during increased cytotoxicity mitochondria were always seen near the nuclear membrane.

Other abnormalities and the degree of impairment in the splenic cells, increased with the increase in concentration and duration of endosulfan administration.

Cells that are damaged by exposure of toxic chemicals undergo a characteristic series of changes. From light microscopic study it is clear that endosulfan induces degeneration followed by apoptosis in spleen of mice. The splenic cells, mainly the blood cells show condense small dark nuclei, which are the specific signs of apoptosis. In endosulfan treated mice, increased level of Reactive Oxygen Species (ROS) have also been observed. Similar signs of cell disintegrity have been suggested by Fredrikssonk, (et al. 2002). Endosulfan stimulates the increased level of oxidants (ROS) within the cell and these oxidants damage the DNA, due to negative signals, received by target cells. Molecules that bind to specific receptors on the cell surface give signal to the cell to begin the apoptosis. These death activators Tumor Necrosis factor - alpha bind to TWF receptor Lymphotoxin. These molecules are more enhanced as blebs that bind to a cell surface receptor. This is a fascinating phenomenon which is evident from electron microscopic study.

Cell proliferation and apoptosis have been reported in many flbroblast and red blood cells. Kerr et al. (1972) Neote et al. (1993) and Fredrikss et al. (2004) also reported proliferation followed by apoptosis in human lung flbroblast. This type of proliferation has never been seen in spleen after chemical injury.

Seong et al. have reported three zones, namely perinuclear, intermediate and marginal in the cytoplasm of the megakaryocyte of rat spleen. The perinuclear zone is characterized by Golgi complex, ribosome, endoplasmic reticulum and mitochondria. The intermediate zone contains a large number of these cell organelles. The marginal zone is almost devoid of cell organelles. In the present investigation all these cell organelles including mitochondria are scattered in the cytoplasm. Their occurrence depends on the cellular activity in normal cell or cellular cytotoxicity in endosulfan treated cells. Phagocytotic activities are abundant during chemo toxicity inducing apoptosis. (Lodish et 2004).

Many workers have reported role of mitochondria in apoptosis. (Budihardjo, 1999; Susin, 1999 Lee, 2000). Authors have also recorded that endosulfan induces permeability transition (PT) of the mitochondrial membrane which constitutes the first rate limiting event of the common pathway of apoptosis. It also disturbs and damages the intervascular pathway. Based on the electron microscopic study, it can be concluded that high chemotoxicity of endosulfan induces mitochondrial degeneration, causing degradation phase of apoptosis i.e. plasma membrane alteration.

ACKNOWLEDGEMENTS

Authors are grateful to all members of SIF-EM unit, AIIMS, New Delhi for their help in Electron Microscopy.

REFERENCES

Barcroft T.J., and Stephens, J.G. 1957. Observations on the size of spleen, J. Physical, Vol. 64, pp. 1-22

Budihardjo, I., et al., Biochemical pathways of cascade activation during apoptosis. Annu. Rev. Cell Dev. Biol., 15, 269-290 (1999).

Crittenden, P.L., Carr, R. and Pruett, S.B. 1998. Immunotoxicological assessment of methyl parathion in Female B6C3F1 mice. J. Toxicol. Environ. Health A. 54 (1): 1-20

Fedriksson K., Lundahi J, Palmberg L et al. Red blood cells stimulate human lung fibroblasts to secrete interleukin-8. Inflammation 2003; 27 (2): 71-8.

Extoxnet Extension of Toxicology Network. USA (1996) A Critique.

Kerr JF, Wyllie AH, Currie A.R. Apoptosis: A basic biological phenomenon with wide-ranging implications in tissue kinetics. Br. J. Cancer 1972; 26 (4): 239-57.

Knisely. M.H. 1936. Spleen Studies. Anat. Rec. Vol. 65, pp. 13-50; 131-148.

Lee, H., and Wei, Y., Mitochondrial role in life and death of the cell. J. Biomed. Sci., 7, 2-15 (2000).

Lodish et al. "In Molecular Cell Biology" Eight Edition, Freeman publication. Page no. 228-229, 924-925.

Nath et al. 2006 endosulfan increases ROS production in the liver of Swiss albino mice. Journal of Eco-physiology and occupational health (communicated).

Neote K, Darbonne W, Ogez J. Hork R, Schall T.J. Identification of a promiscuous inflammatory peptide receptor on the surface of red blood cells, J. Biol. Chem. 1993; 268 (17): 12247-9.

P.R. Wheater H.G. Burkitt, V.G. Daniels . Functional histology, Second Edition. A Text and Colour Atlas.

Peck, H.M., and Hoerr, N.L. 1951. The intermediate circulation in the red pulp of mouse spleen. Anat. Rec. Vol. 109, pp. 447. 478.

Seema Rai 2002, Effect of organophosphorous compound on spleen of Mice. P.U. Thesis.

Seong S. Han, Burton L, Baker. The anatomical record, volume 149, Issue 2, pp. 251-267 Published online : 3 Feb 2005.

Susin, S.A., et al., Molecular characterization of mitochondrial apoptosis-inducing factor. Nature, 397, 441-446 (1999).

Voccia, Blackley, B; Brousseau P. and Fourmier, M. (1999) Immunotoxicity of pesticides. A review, Toxicology and Industrial Health, 15, 119-132.

Wojdani, A. and Attaradeh M. et al. 1984. Immunocytotoxicity effects of polycyclic aromatic hydrocarbons of mouse lymphocytes. Toxicol. 31 : 181-189.

Yoffey, J.M. (editor) 1967. Bristol Symposium : The lymphocyte in Immunology and Haemopoesis. The William and Willikins Company Baltimore.

Comparison of Natural and Synthetic Insecticides on the Incidence of few Major Pests of Okra

CHANDRAWATI JEE AND BINDU SINGH***

SUMMARY

Insect pest infestation and incidence are the most limiting factor for accelerating yield potential. The major insect pests of Orka includes leaf hopper (Jassids) shoot and fruit borer, leaf roller, red cotton bug, red spider mite and many minor insects pests. Out of thirteen major insect pests at different stages of growth, Jassid *(Amrascas biguttula, ishida)* and shoot/ fruit borer *(Earias vittella iabricius)* are considered very destructive and serious pests and damage the plants and reduces the yield. The damage yield potential of both insects was studied during summer season of 2002 and 2003 at A.N. College, experimental fields. Comparative studies of natural and synthetic pesticides were studied and it was recorded that natural pesticide showed the best response both in terms of tolerance and yield potential of the crop with the maximum profit margin.

Key words: Okra, Incidence, yield potential, natural and synthetic insecticides.

Introduction: Okra crop is attacked by many minor and major pests at various stage of development. Few of them are very destructive and damage potential was very high. As per an estimate approximate 50 to 80% fruit yield losses due to infection and infestation of Jassids and borer pests and reducing

* Department of Biotechnogy, A.N. College, Patna.

** Department of Zoology, College of Commerce, Patna.

the margin of growers. Several synthetic insecticides have been effective against these pests but they are not eco-friendly and create the problems like

(a) Development of resistance to the insects

(b) Hazardous effects on the environment

(c) Residual toxicity in vegetative and in soil

(d) Create health hazard for consumers

(e) Develop resistance tolerance among insects/pests

In view of the ill effects of the organic synthetic pesticides in long run including pest resurgence, pest resistance, pesticides residues in foods, pollution of water, soil and atmosphere. High cost and hazards involved in their usage, safer alternative of crop protection such as use of natural plant product as insecticides/pesticides is gaining importance. These natural botanical pesticides are easily biodegradable and offer immense scope in integrated pest management. Here different products of neem were used with pest management in okra crop.

Therefore a comparative study of natural and synthetic insecticides among different formulation was studied and the yield potentiality of Okra was calculated and recorded. Many researchers worked on the use of botanicals for management of insects in vegetable. Sundra et al., 1984, Singh et al., 1985, Mohan; 1989, Saxena; 1989 Kathirvel; 1988, Srinivas; 1990, Schmuttererererer et al., 1990 described bioactivity of Neem against insects. Neem has been evaluated against 305 species of insects belonging to 10 orders. Most of the above evaluations deal with pests of agricultural importance and include almost all the key pests of agriculture.

MATERIAL & METHODS

The experiment was done in plots at the A.N. College, Patna during two successive years (2002 & 2003) in randomized block design with four replicates. The plot size is 3 m x 2 m with plant spacing of 30 cm x 30 cm. Certified seeds (Pusa Sawani) were sown on the fields on the same date in both years of experimentation and recommended package of practice was adopted.

Natural and synthetic insecticidal treatments were applied with a Maruti Hand Pump Sprayer with the pressure created by 40 strokes of air pump. Uniform spraying was done. Different concentration of insecticides was given in constant water 1000 litre/ ha at 20 days intervals. The incidence of Jassid was recorded at weekly intervals on 30 leaves /10 plants from the time of pest appearance upto the harvest of fruit per replicate. Average of 10 randomly selected tagged plants in each replicate was recorded. Population count was taken together and average population 710 plants / 30 leaves were calculated in both the years. Pooled data were selected for analysis of variance. The incidences of shoot / fruit borer were recorded periodically and mean value of incidence was calculated by unitary method by using formula.

(a) % Shoot Damage =

$$\frac{\text{No. of damage shoot}}{\text{Total no. of shoots (healthy and damage)}} \times 100$$

(b) % Fruit damage by number =

$$\frac{\text{No. of damage fruit}}{\text{Total No. of fruit ((healthy and damage)}} \times 100$$

(c) % fruit damage by weight =

$$\frac{\text{Wt. of damaged fruit}}{\text{Total wt of fruit ((healthy and damage)}} \times 100$$

After each picking, the larval population was counted in every hundred fruits. The pooled data of each treatment were recorded.

RESULTS & DISCUSSION

During the study period (2002-2003) an extensive survey and surveillance for the insect pests of Okra crop was conducted on regular interval to know the incidence and population dynamics. A total number of 13 insect pests were infesting the Okra crop in experimental area damaging the crop to various extents. Two major and serious insect pests Jassids and fruit / shoot borer pest were selected for investigation because they cause severe damage to the crop.

Table 21.1: Effect of foliar application of Neem product and Carbonyl in control of jassids in Okra crop (pooled mean 2002 & 2003)

Sl. No.	*Treatment*	*Mean jassid population 730 leaves*	*Mean % of shoot infestation*	*Mean % of fruit infestation by*		*Mean No. of larva/ 100 fruit*	*Av. marketable fruit yield q/ha^{-1}*
				Number	Weight		
1.	T_1 Control	15.10	30.03	33.75	32.11	49.90	77.90
2.	T_2 (NSWE)	10.14	14.27*	14.32*	102.86	22.06	108.38
3.	T_3 (Neem oil)	9.09	15.13*	16.43*	23.12	21.23	115.52
4.	T_4 (Carbonyl)	10.52	16.25*	18.97*	20.12	25.67	102.61

Significant at 5% level.

Table 21.2: Result of trial jassids on Okra Crop (2002-2003)

Sl. No.	*Treatment*	*Damage leaves count/ extent of damage (0-3)*				
		Count before spray	*Count 7 days*	*Count 15 days*	*Count 7 days*	*Count 15 days*
1.	T_1 Control	0	1	2	3	3
2.	T_2 (NSWE)	0	1	1	1	2
3.	T_3 (Neem oil)	0	0	1	1	1
4.	T_4 (Carbonyl)	0	0	1	1	1

* 0 to 3 indicate extent of damage.

Table 21.3: Economics of natural and synthetics pesticides on incidence of pest infestation.

SI. No.	*Treatment*	*Marketable fruit yield q/ha^{-1}*	*Increased in yield over control*	*Gross Income Rs./ha^{1}*	*Cost of treatment*	*Net Income Rs./ha^{1}*
1.	T_1 Control	77.90	—	—	—	~
2.	T_2 (NSWE)	108.38	20.48	12800.00	1281.50	11518.50
3.	T_3 (Neem oil)	115.52	27.62	17262.50	1431.50	15831.00
4.	T_4 (Carbonyl)	102.61	14.71	9193.75	27.95	6397.80

The results of the experiments conducted in the field with okra crop for assessment of effectiveness of neem extracts along with carbonyl 0.5% and untreated control was presented in Table 21.I. The observations were taken by counting damaged leave of 10 plants from the middle lower and upper portion of the plants. The counting was classified on the basis of severity of damage i.e. if no damage was found in 10 plants counted it was classified as zero (0) and accordingly to the extent of damage 1, 2 and 3 was given Table 21.2.

The result shows that the application of NSWE (Neem Seed Water Extract) and Neem oil was very effective in checking the effluent of damage and population density of the jassid statistically. It was observed that NSWE and neem oil were more effective than carbonyl in controlling the pest. The efficacy of NSWE and neem oil is at par. Similar results have been found by other workers working with other vegetable crops Singh et al 1993. Result indicated that Jassid population varied from 9.09 to 15.10/30 leaves. It was observed that N. oil was very effective in Jassid population control while NSWE and 0.5% carbonyl were at par in controlling the population. 0.5% carbonyl is responsible for increasing pest resistance.

On the basis of pooled means of shoot infestation of shoot and fruit borer and mean Jassid population it was observed that all the treatments were significantly superior over control. The minimum shoot infestation (14.27%) was recorded in treatment T_2 while maximum shoot infestation (30.03%) was recorded in T_1, T_3 and T_4 were at par in controlling the shoot infestation. The pooled mean of fruit infestation revealed that treatment T_2 recorded the minimum fruit infestation 14.32 and 17.86% both on fruit number and weight respectively. However it is followed by T_3 (16.43 & 23.12%) and T_4 (18.97 & 20.12%) both on fruit number and fruit weight respectively.

The pooled means of the larval population indifferent treatments varied from 21.23 to 49.90/100 fruits. The minimum larval population of 21.23/100 fruits was found in T_3 followed by T_2 (22.06/100 fruits) and T_4 25.67/100 fruits. The larval population treatment T_3 and T_2 were recorded at par indicating the NSWE and Neem oil both are effective in controlling larval population without any side effect. Result indicated that synthetic pesticide T_4 (0.5% carbonyl) was less effective than the natural pesticides in controlling infestation of the borer pest on fruit number and weight as well as larval population.

The yields were significantly higher (102.61 to 115.52 qha^{-1}) over control 77.90 q ha^{-1}. The neem oil combination recorded the higher fruit yield (115.52 q ha^{-1}) followed by T_2 (108.30 q ha^{-1}) and T_4 (102.61 q ha^{-1}) (Table 21.1). Result indicated that yield of Okra crop significantly increased after application of synthetics and natural pesticides. However, synthetic pesticides have some side effect where as natural pesticides eco-friendly.

By the perusal of Table 21.3 it was clear that after application of pesticides (both natural and synthetics) the increase in yield was recorded. Maximum gross income Rs. 17262.50/ha was recorded in treatment T_3 while the maximum net income Rs. 15831.00/ha was recorded in same treatment. Synthetic pesticides and their application need higher investment cost therefore the profit margin was less in T_4.

On the basis of above finding the application of balanced dose of pesticides significantly reduced the incidence of Jassid and borer pest and help in increasing the yield of Okra. When economics of plant protection is considered the control of Okra pest with natural pesticides is much profitable than synthetic pesticides and also help in preserving the environment and ecosystem without any residual toxicity. Pests can not develop resistance with natural pesticides and will also help in conserving the natural enemies of the pests of Okra. The judicious use of natural pesticides would play a purposeful role as its use is helpful in plant and environment protection and sustainable development.

Market price of Okra healthy fruit was 625/q during 2002-03 at local market.

ACKNOWLEDGEMENT

The authors are thankful to Principal A.N. College for providing field and laboratory for experimentation.

REFERENCES

Kathirvel, M. (1988), Studies on the management of pest and nematodes of bhendi with botanical and insecticides, M.Sc. (Ag.) Thesis TNAU Coimbatore 135 p.

Mohan, K. (1989), Studies on the effect of neem product and vegetable oil against major pests of rice and safety to natural enemies. M.Sc. (Ag) Thesis TNAU, Coimbatore, 134 p.

Saxena, R.C. (1989), Insecticides from neem tree. Insecticides of plant origin pp. 110-135. (Eds. J.T. Arnason, B.J. Rephilogene and Peter Morand) and symposium series 387. American Chemical Society, Washington DC, USA.

Schmuttererer, H.K.R.S. Asher (eds) (1990), properties and potential of natural pesticides from the neem tree, *Azadirachta indica* A. Rev. Ent. 35, 271-297.

Singh, Y.P.; H.S. Bagha, V.K. Vijjan (1985), Toxicity of water Extract of Neem *(Azadirachta indica* A. Juss). Berries in poultry birds, Neem News Letter, Vol. 2; No. 2, pp. 17-18.

Srinivasan K. (1990), Pest management in vegetable crops-invistas. Rahuri in seminar on recent Advances in vegetable Pest Management.

Sundra Babu, P.C. and Rajsekaron, B. (1984), Evaluation of certain synthetic prethroids and vegetable products for the control of benglgram pod borer, *Heliothes armigera* Hibner, pesticides 13: 58-59.

Bibliography

American Association on Mental Deficiency (1973) as cited by Kisker, George W., *The Disorganized Personality,* McGraw Hill (International student Edition-III), 1964.

Andrews, T.G. (Ed.); *Methods of Psychology,* New York: John Wiley, 1960.

Athawale, M.C. (1963). Estimation of height from lengths of forearm bones, a study of one hundred Maharashtrian male adults of ages between twenty five and thirty years.

Am. J. Phys. Anthropol., 21 : 105-112.

Barton, Hall, *Psychiatric Examination of the School Child,* London: Edward Arnold, 1947.

Bingham, W.V.D., *Aptitude and Attitude Testing,* New York: Harper and Brothers, 1937.

Biggie, M.L. and Hunt, M.P. ; *'Psychological Foundations of Education'.* New York: Harper and Row, 1968.

Boring E.C., Lang field, H.S. and Weld, H.P. (ed.); *Foundations of Psychology,* New York.

Brown, F.J. ; *Educational Sociology,* New York. Prentice Hall, 1960 (5th printing).

British Mental Deficiently Act (1929) as cited by Shanmugam, T.E. : *Abnormal Psychology,* New Delhi: Tata McGraw Hill, 1981.

Brown, J.F.; *Educational Sociology,* New York: Prentice Hall, 1960 (5th Printing).

Brown, J.F.; *The Psychodynamics of Abnormal Behavior,* New Delhi: Asia Publishing House (Indian Reprint), 1969.

Bajaj, I.D., Bhardwaj, O.P. and Bhardwaj, S. 1 967). Appearance and fusion of important ossification centres. A study of Delhi population. Indian J. Med. Res., 55: 1064-1067.

Berry, A.C. (1975). Factors affecting the incidence of non-metrical skeletal variants. J. Anal. 120: 519-535.

Brahma, K.C. and Mitra, N.L. (1973). Ossification of carpal bones. A radiological study in the tribals of Chotanagpur. J. Anal. Soc. India, 22 : 21-28.

Charnalia, V.M. (1961). Anthropological study of the foot and its relationship to stature in different castes and tribes of Pond cherry State. J. Anal. Soc. India, 10: 26-30.

Chhatrapati, D.N. and Misra, B.D. (1967). Position of the nutrient foramen on the shafts of human long bones. J. Anal. Soc. India, 16: 1-10.

Das Gupta, S.M., Prasad, V. and Singh, S. (1974). A roentgenologic study of epiphysial union around elbow, wrist and knee joints and the pelvis in boys and girls of Uttar Pradesh. J. Indian Med. Assoc., 62 : 10-12.

Garg, K., Bahl, I., Mathur, R. and Verma, R.K. (1985). Bilateral asymmetry in surfacearea of carpal bones in children. J. Anal. Soc. Ind. 34: 167-172.

Garg, K., Kulshrestha, V., Bahl, I., Mathur, R., Mittal, S.K. and Verma, R. (1988).

Asymmetry in surface area of capitate and hamate bones in normal and malnourished children. J. Anal. Soc. Ind. 37: 73-74.

Gupta, A., Garg, K., Mathur, R. and Mittal, S.K. (1988). Asymmetry in surface area of bony epiphysis of knee in normal and malnourished children. Ind. Paed. No. 10 : 999-1000.

Halim, A. and Siddiqui M.S. (1976). Sexing of atlas, Indian J. Phys. Anthropol. Hum.Genet., 2 : 129-135.

Hasan. M. and Narayan. D. (1963). The ossification centres of carpal bones. A radiological study of the times of appearance in V.P. Indian subjects. Indian J.Med. Res., 51 : 917-920.

Hasan. M. and Narayan, D. (1964). Radiological study of the postnatal ossification of the upper end of humerus in V.P. Indians. J. Anat. Soc. India. 13: 70-75.

Jeyasingh, P., Saxena, S.K., Arora, A.K., Pandey, D.N. and Gupta, C.D. (1980). A study of correlations between some neonatal and placental parameters. J. Anat. Soc.India, 29: 14-18.

Jet, I. (1957). Observations on prenatal ossification with special reference to the bones of hand and foot. J. Anat. Soc. India, 6: 12-23.

Jit, I. (1972). Times of ossification in India (editorial). Bull. PGI (Chandigarh), 6: 100-103.

Jit. I. (1979). The house of skeletons case. J. Allat. Soc. India. 28: 106-116.

Jit, I. Hinging, V. and Kulkarni. M. (1980). Sexing of human sternum. Am. J. Phys. Anthropol. 53 : 217-224.

Jit. I. and Gandhi, O P. (1966). The value of preauricular sulcus in sexing bony pelvis. J. Anat. Soc. India, 15 : 104-107.

Jit, I. and Kulkarni, M. (1976). Times of appearance and fusion of epiphysis at the medial end of the clavicle. Indian J. Med. Res.. 64: 773-782.

Jit, I. and Singh. S. (1956). Estimation of the stature from the clavicle. Indian J. Med.Res. 44 : 133-135.

Jit. I. and Singh S. (1966). Sexing of the adult clavicle. Indian J. Med. Res., 54: 55 1-571.

Jit. I. and Singh. B. (1971). A radiological study of the time of fusion of certain epiphyses in Punjabees. J. Anat. co. India. 20: 1-27.

Jit, I.. Verma, V. and Gandhi, O.P. (1968). Ossification of bones of the hand and foot in newborn children. J. Anat. Soc. India, 17; 8-13.

Kate, B.R. (1964). A study of the regional variation of the Indian femur, the diameter of the head, its medicolegal and surgical application. J. Anat. Soc. India, 13 : 80-84.

Kate. B.R (1967). The angle of the femoral neck in Indians. East. Anthropologist. 20 : 54-60.

Kate, B.R. (1971). Nutrient foramina in human long bones. J. Anat. Soc. India, 20 : 139-144.

Kate. B.R. and Majumdar, R.D. (1976). Stature estimation from femur and humerus by regression and autometry. Acta Anat.. 94 : 311-320.

Kolte. P.M. and Bansal, P.C. (1974). Determination of regression formulae for reconstruction of stature from the long bones of upper limb in Maharashtrians of Marathwada region. J. Anat. Soc. India, 23: 6-11.

Krogman, W.M. (1962). The Human Skeleton in Forensic Medicine. Thomas. Springfield, 111.

Longia, G.S. Ajmani, M.L., Saxena, S.K. and Thomas, R.J. (1980). Study of diaphysial nutrient foramina in the human long bones. Acta Anat., 107 : 399-406.

Maniar. B.M. Seervai, M.H. and Kapur. P.L. (1974). A study of ossification centres in the hand and wrist of Indian children. Indian Pediatr. 1:203-211.

Modi, N.J. (1977). Modi's Textbook of Medical Jurisprudence and Toxicology. 20th ed., p. 27-38. and 80-88. Tripathi. Bombay.

Mysorekar, V.R. (1967). Diaphysial nutrient foramina in human long funes. J. Anal,101 : 813-822.

Mysorekar, V.R. and Nandedkar, A.N. (1979). Diaphysial nutrient foramina in human phalanges. J. Anal. 128:315-322.

Mysorekar, V.R., Verma, P.K., Nandedkar, A.N. and Sarma, T.C.S.R. (1980). Estimation of stature from parts of bones, lower end of femur and upper end of radius. Medicine Science anti Law Lond.20: 283-286.

Notes

Notes

Notes

Notes